Power Line Worker

Level Two: Transmission

Trainee Guide

Prentice Hall

Boston Columbus Indianapolis New York San Francisco Upper Saddle River
Amsterdam Cape Town Dubai London Madrid Milan Munich Paris Montreal Toronto
Delhi Mexico City São Paulo Sydney Hong Kong Seoul Singapore Taipei Tokyo

National Center for Construction Education and Research
President: Don Whyte
Director of Product Development: Daniele Stacey
Power Line Worker Project Manager for Transmission: Matt Tischler
Production Manager: Tim Davis
Quality Assurance Coordinator: Debie Ness
Desktop Publishing Coordinator: James McKay
Production Specialist: Heather Griffith-Gatson
Editor: Chris Wilson

Writing and development services provided by Topaz Publications, Liverpool, NY
Lead Writer/Project Manager: Thomas Burke
Desktop Publisher: Joanne Hart
Art Director: Megan Paye
Permissions Editors: Andrea LaBarge, Alison Richmond
Writers: Thomas Burke, Troy Staton, Gerald Shannon, Patricia Vidler

Pearson Education, Inc.
Editorial Director: Vernon R. Anthony
Executive Editor: Alli Gentile
Senior Product Manager: Lori Cowen
Operations Supervisor: Deidra M. Skahill
Art Director: Jayne Conte
Director of Marketing: David Gesell
Executive Marketing Manager: Derril Trakalo
Marketing Manager: Brian Hoehl
Marketing Coordinator: Crystal Gonzalez

Composition: NCCER
Printer/Binder: Document Technology Resources, Fredericksburg, VA
Cover Printer: Document Technology Resources, Fredericksburg, VA
Text Fonts: Palatino and Univers

10 9 8 7 6 5 4 3 2

PEARSON

www.pearsonhighered.com

ISBN-13: 978-0-13-273033-4
ISBN-10: 0-13-273033-2

Preface

To the Trainee

Welcome to your second year of training in power line work. If you are training under an NCCER Accredited Training Program Sponsor, you have successfully completed *Power Industry Fundamentals* and *Power Line Worker Level One*, and are well on your way to more advanced training.

NCCER's *Power Line Worker Level Two: Transmission* covers the skills and knowledge necessary to develop outstanding transmission line workers. Topics addressed include three-phase systems, structure rigging and erection, equipment installation, and system maintenance. By the end of this training, you will have learned how to properly and safely erect transmission structures, pull line, use monitoring equipment, and repair line damage. You will also be able to apply electrical theory to your everyday duties.

As our population grows, the demand for electricity also increases. Transmission lines are necessary to address this need and power line workers have the skills required to install and maintain our expanding energy infrastructure.

Many opportunities are available for well-trained line workers, and employment can be found in the fields of natural gas distribution, power generation, and government. Power line workers are expected to be in high demand, as more than 20% of the current workforce will retire by the next decade. And the demand will only intensify as the nation struggles to modernize an aging energy infrastructure.

We wish you success as you progress through this training program. Should you have any comments on how NCCER might improve upon this textbook, please complete the User Update form located at the back of each module and send it to us. We will always consider and respond to input from our customers.

We invite you to visit NCCER's website at **www.nccer.org** for information on the latest product releases and training, as well as online versions of the *Cornerstone* newsletter and Pearson's product catalog.

Your feedback is welcome. You may email your comments to **curriculum@nccer.org** or send general comments and inquiries to **info@nccer.org**.

NCCER Curricula

NCCER is a not-for-profit 501(c)(3) education foundation established in 1996 by the world's largest and most progressive construction companies and national construction associations. It was founded to address the severe workforce shortage facing the industry and to develop a standardized training process and curricula. Today, NCCER is supported by hundreds of leading construction and maintenance companies, manufacturers, and national associations. The NCCER Curricula was developed by NCCER in partnership with Pearson Education, Inc., the world's largest educational publisher.

Some features of NCCER's Curricula are as follows:

- An industry-proven record of success
- Curricula developed by the industry for the industry
- National standardization providing portability of learned job skills and educational credits
- Compliance with the Office of Apprenticeship requirements for related classroom training *(CFR 29:29)*
- Well-illustrated, up-to-date, and practical information

NCCER also maintains a National Registry that provides transcripts, certificates, and wallet cards to individuals who have successfully completed modules of NCCER's Curricula. *Training programs must be delivered by an NCCER Accredited Training Sponsor in order to receive these credentials.*

Special Features

In an effort to provide a comprehensive, user-friendly training resource, we have incorporated many different features for your use. Whether you are a visual or hands-on learner, this book will provide you with the proper tools to get started in the power line worker industry.

Introduction

This page is found at the beginning of each module and lists the Objectives, Performance Tasks, Trade Terms, and Required Trainee Materials for that module. The Objectives list the skills and knowledge you will need in order to complete the module successfully. The Performance Tasks give you the opportunity to apply your knowledge to the real world duties that power line workers perform. The list of Trade Terms identifies important terms you will need to know by the end of the module. Required Trainee Materials list the materials and supplies needed for the module.

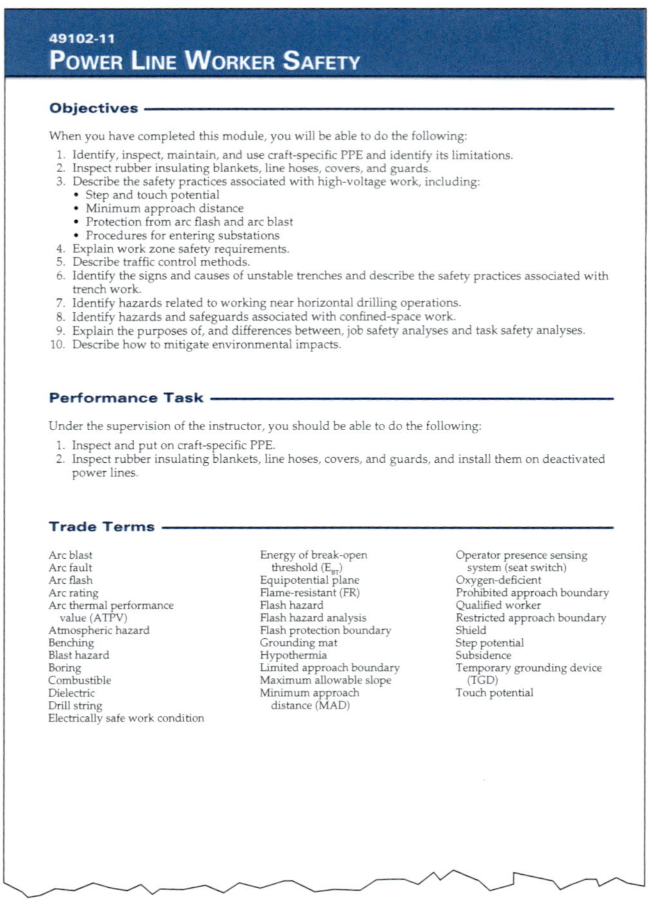

49102-11
POWER LINE WORKER SAFETY

Objectives

When you have completed this module, you will be able to do the following:

1. Identify, inspect, maintain, and use craft-specific PPE and identify its limitations.
2. Inspect rubber insulating blankets, line hoses, covers, and guards.
3. Describe the safety practices associated with high-voltage work, including:
   - Step and touch potential
   - Minimum approach distance
   - Protection from arc flash and arc blast
   - Procedures for entering substations
4. Explain work zone safety requirements.
5. Describe traffic control methods.
6. Identify the signs and causes of unstable trenches and describe the safety practices associated with trench work.
7. Identify hazards related to working near horizontal drilling operations.
8. Identify hazards and safeguards associated with confined-space work.
9. Explain the purposes of, and differences between, job safety analyses and task safety analyses.
10. Describe how to mitigate environmental impacts.

Performance Task

Under the supervision of the instructor, you should be able to do the following:

1. Inspect and put on craft-specific PPE.
2. Inspect rubber insulating blankets, line hoses, covers, and guards, and install them on deactivated power lines.

Trade Terms

Arc blast	Energy of break-open threshold (E_{BT})	Operator presence sensing system (seat switch)
Arc fault	Equipotential plane	Oxygen-deficient
Arc flash	Flame-resistant (FR)	Prohibited approach boundary
Arc rating	Flash hazard	Qualified worker
Arc thermal performance value (ATPV)	Flash hazard analysis	Restricted approach boundary
Atmospheric hazard	Flash protection boundary	Shield
Benching	Grounding mat	Step potential
Blast hazard	Hypothermia	Subsidence
Boring	Limited approach boundary	Temporary grounding device (TGD)
Combustible	Maximum allowable slope	Touch potential
Dielectric	Minimum approach distance (MAD)	
Drill string		
Electrically safe work condition		

Color Illustrations and Photographs

Full-color illustrations and photographs are used throughout each module to provide vivid detail. These figures highlight important concepts from the text and provide clarity for complex instructions. Each figure reference is denoted in the text in *italic type* for easy reference.

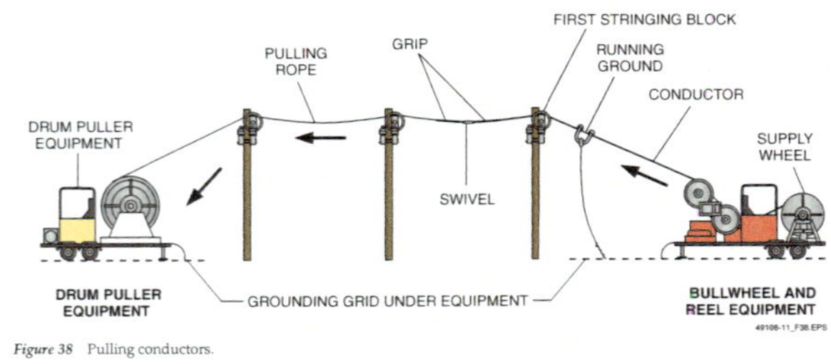

Notes, Cautions, and Warnings

Safety features are set off from the main text in highlighted boxes and are organized into three categories based on the potential danger of the issue being addressed. Notes simply provide additional information on the topic area. Cautions alert you of a danger that does not present potential injury but may cause damage to equipment. Warnings stress a potentially dangerous situation that may cause injury to you or a co-worker.

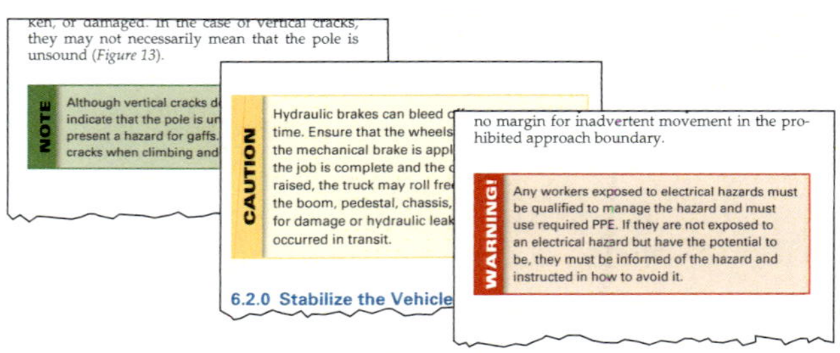

Did You Know?

The Did You Know? features offer hints, tips, and other helpful bits of information from the trade.

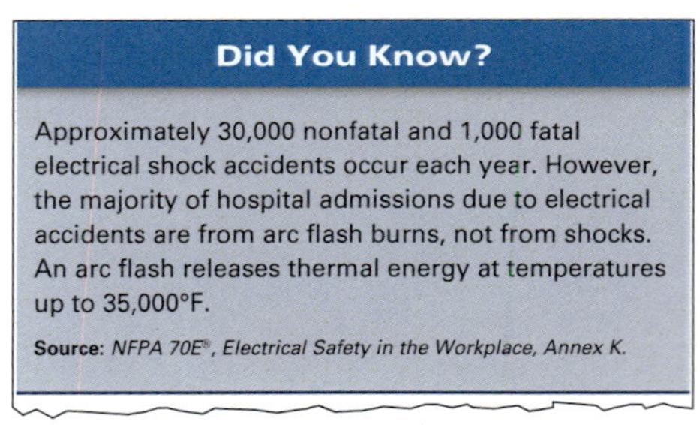

On Site

On Site features provide a head start for those entering the electrical transmission and distribution fields by presenting technical tips and professional practices from power line workers on a variety of topics. On Site features often include real-life scenarios similar to those you might encounter on the job site.

On Site

Transformers

Large distribution transformers at power substations step down the power to the level required for local distribution. Pole transformers, such as the one shown here, step the power down further to the voltages needed for homes and businesses.

49103-11_SA02.EPS

Think About It

Think About It features use "What if?" questions to help you apply theory to real-world experiences and put your ideas into action.

Think About It

The Magic of Electricity

The effect of the flow of electrons occurs at a speed that is close to the speed of light, about 186,000 miles per second. How long does it take the light from the end of a flashlight to reach the floor? If you ran a light circuit from Maine to California and flipped the switch, how long would it take for the light to come on?

Step-by-Step Instructions

Step-by-step instructions are used throughout to guide you through technical procedures and tasks from start to finish. These steps show you not only how to perform a task but also how to do it safely and efficiently.

trically safe work condition. Most companies have detailed written procedures for performing this task.

Step 1 Determine whether any other crews are working on the circuit. All distribution lines are treated as if they are energized unless your team has performed a de-energizing procedure. If two or more crews are working on the same lines or equipment, each crew must independently perform a de-energizing procedure and apply their own lockout/tagout devices to energy controls.

Step 2 Designate one qualified member of the crew as the employee in charge of the electrical clearance.

Trade Terms

Each module presents a list of Trade Terms that are discussed within the text and defined in the Glossary at the end of the module. These terms are denoted in the text with **bold, blue type** upon their first occurrence. To make searches for key information easier, a comprehensive Glossary of Trade Terms from all modules is located at the back of this book.

You will hear the term *circuit* throughout your training. An electrical circuit contains, at minimum, a **voltage** source, a load, and **conductors** (wires) to carry the electrical **current** (*Figure 1*). The circuit should also have a means to stop and start the current, such as a switch.

Electricity is concerned with cause and effect. The presence of voltage (volts) in a closed circuit causes current (amps) to flow. The more voltage you apply, the more current will flow. However, the amount of current flow is also determined by how much **resistance**, in **ohms** (Ω), the load offers

Review Questions

Review Questions are provided to reinforce the knowledge you have gained. This makes them a useful tool for measuring what you have learned.

Review Questions

1. All of the following items are considered by OSHA to be acceptable fall prevention systems for working at heights above 6 feet *except* for _____.
   a. a guardrail system
   b. personal fall protection
   c. the two-climber buddy system
   d. a safety net system
2. Fall arrest systems _____.
   a. stop or take control of a fall in progress
   b. prevent falls from ever occurring
   c. ensure that the climber does not swing from side-to-side during a fall
   d. ensure that every fall is reported to the proper authorities
3. OSHA requires that a PFAS limits the maximum force imparted to the body during fall arrest to _____.
   a. 500 pounds
   b. 1,200 pounds
   c. 1,800 pounds
   d. 2,500 po

7. When an anchor point rated at 5,000 pounds is not available, the climber must then ensure that the chosen anchor position can at least withstand the maximum load that could be placed on it during a fall arrest, multiplied by a factor of _____.
   a. 1.5
   b. 2
   c. 3
   d. 5
8. Anchor points that are used as fall restraint anchors, and *not* as fall arrest anchors, must be load rated to handle _____.
   a. 2,500 pounds
   b. 3,000 pounds
   c. 5,000 pounds
   d. 7,500 pounds
9. The climber's personal fall arrest harness must be connected to the anchor point using _____.
   a. at least two lanyards

NCCER Curricula

NCCER's training programs comprise more than 80 construction, maintenance, pipeline, and utility areas and include skills assessments, safety training, and management education.

Boilermaking
Cabinetmaking
Carpentry
Concrete Finishing
Construction Craft Laborer
Construction Technology
Core Curriculum:
  Introductory Craft Skills
Drywall
Electrical
Electronic Systems Technician
Heating, Ventilating, and
  Air Conditioning
Heavy Equipment Operations
Highway/Heavy Construction
Hydroblasting
Industrial Coating and Lining
  Application Specialist
Industrial Maintenance
  Electrical and Instrumentation
  Technician
Industrial Maintenance
  Mechanic
Instrumentation
Insulating
Ironworking
Masonry
Millwright
Mobile Crane Operations
Painting
Painting, Industrial
Pipefitting
Pipelayer
Plumbing
Reinforcing Ironwork
Rigging
Scaffolding
Sheet Metal
Signal Person
Site Layout
Sprinkler Fitting
Tower Crane Operator
Welding

Green/Sustainable Construction

Building Auditor
Fundamentals of
  Weatherization
Introduction to Weatherization
Sustainable Construction
  Supervisor
Weatherization Crew Chief
Weatherization Technician
Your Role in the Green
  Environment

Energy

Alternative Energy
Introduction to the Power
  Industry
Introduction to Solar
  Photovoltaics
Introduction to Wind Energy
Power Industry Fundamentals
Power Generation Maintenance
  Electrician
Power Generation I&C
  Maintenance Technician
Power Generation Maintenance
  Mechanic
Power Line Worker
Solar Photovoltaic Systems
  Installer
Wind Turbine Maintenance
  Technician

Pipeline

Control Center Operations,
  Liquid
Corrosion Control
Electrical and Instrumentation
Field Operations, Liquid
Field Operations, Gas
Maintenance
Mechanical

Safety

Field Safety
Safety Orientation
Safety Technology

Management

Fundamentals of Crew
  Leadership
Project Management
Project Supervision

Supplemental Titles

Applied Construction Math
Careers in Construction
Tools for Success

Spanish Translations

Basic Rigging
  (Principios Básicos de
  Maniobras)

Carpentry Fundamentals
  (Introducción a la Carpintería,
  Nivel Uno)

Carpentry Forms
  (Formas para Carpintería,
  Nivel Trés)

Concrete Finishing, Level One
  (Acabado de Concreto,
  Nivel Uno)

Core Curriculum:
  Introductory Craft Skills
  (Currículo Básico:
  Habilidades Introductorias
  del Oficio)

Drywall, Level One
  (Paneles de Yeso, Nivel Uno)

Electrical, Level One
  (Electricidad, Nivel Uno)

Field Safety
  (Seguridad de Campo)

Insulating, Level One
  (Aislamiento, Nivel Uno)

Ironworking, Level One
  (Herrería, Nivel Uno)

Masonry, Level One
  (Albañilería, Nivel Uno)

Pipefitting, Level One
  (Instalación de Tubería
  Industrial, Nivel Uno)

Reinforcing Ironwork, Level One
  (Herreria de Refuerzo,
  Nivel Uno)

Safety Orientation
  (Orientación de Seguridad)

Scaffolding
  (Andamios)

Sprinkler Fitting, Level One
  (Instalación de Rociadores,
  Nivel Uno)

Acknowledgments

This curriculum was revised as a result of the farsightedness and leadership of the following sponsors:

Baltimore Gas and Electric
Constellation Energy
MasTec, Inc.
Pike Electric Corporation
Quanta Services Inc.

This curriculum would not exist were it not for the dedication and unselfish energy of those volunteers who served on the Authoring Team. A sincere thanks is extended to the following:

David Brzozowski
Don Daniel
Robert Groner
Gordon Johnson
David Kammer
David McPeak
Kevin Watson
Terry Williams

NCCER Partners

American Fire Sprinkler Association
Associated Builders and Contractors, Inc.
Associated General Contractors of America
Association for Career and Technical Education
Association for Skilled and Technical Sciences
Carolinas AGC, Inc.
Carolinas Electrical Contractors Association
Center for the Improvement of Construction Management and Processes
Construction Industry Institute
Construction Users Roundtable
Construction Workforce Development Center
Design Build Institute of America
Merit Contractors Association of Canada
Metal Building Manufacturers Association
NACE International
National Association of Minority Contractors
National Association of Women in Construction
National Insulation Association
National Ready Mixed Concrete Association
National Technical Honor Society
National Utility Contractors Association
NAWIC Education Foundation
North American Technician Excellence
Painting & Decorating Contractors of America
Portland Cement Association
SkillsUSA
Steel Erectors Association of America
The Manufacturers Institute
U.S. Army Corps of Engineers
University of Florida, M.E. Rinker School of Building Construction
Women Construction Owners & Executives, USA

NCCER Business Partners

Contents

Module Five

Transmission System Maintenance

Coverage includes safety practices related to working with helicopters, as well as inspection of insulators, towers, and poles. The module covers clearance procedures, as well as environmental concerns such as protection of wetlands, waterways, and wildlife. (Module ID 81204-11; 40 Hours)

Glossary

Index

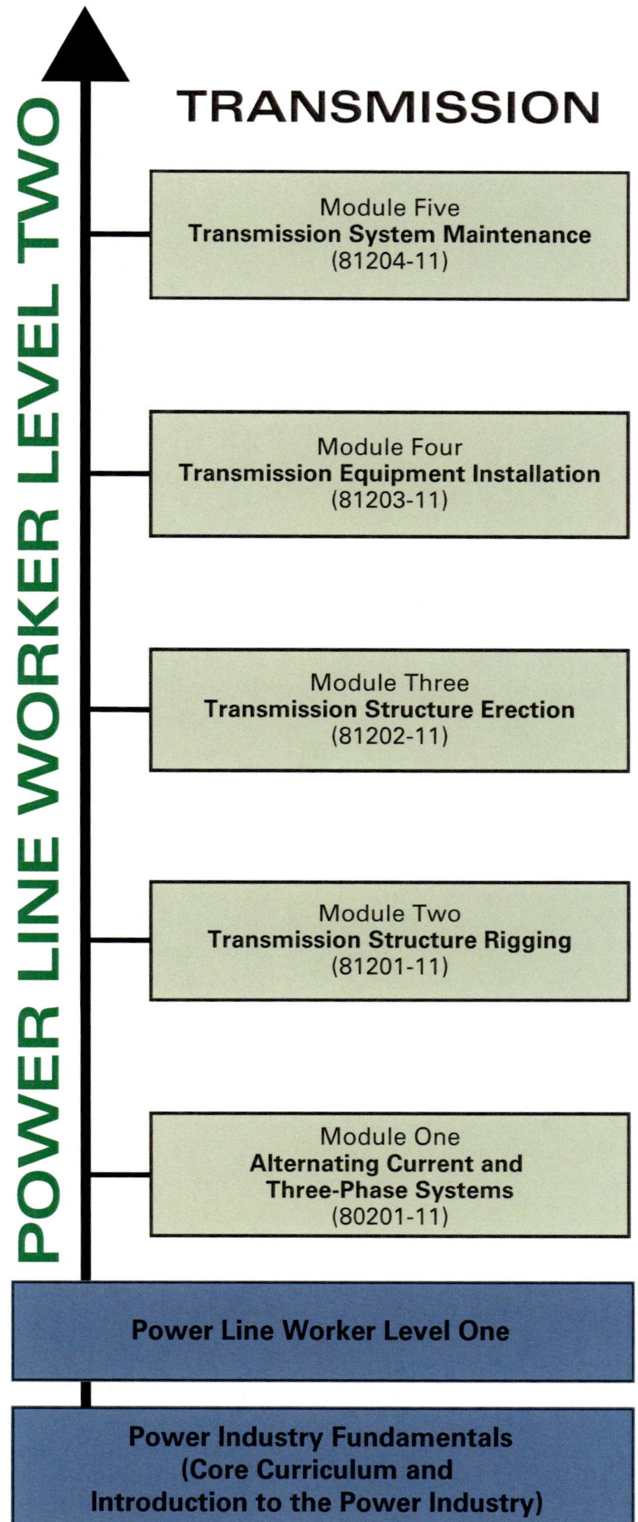

This course map shows all of the modules in *Power Line Worker Level Two: Transmission*. The suggested training order begins at the bottom and proceeds up. Skill levels increase as you advance on the course map. The local Training Program Sponsor may adjust the training order.

80201-11

Alternating Current and Three-Phase Systems

Module One

Trainees with successful module completions may be eligible for credentialing through NCCER's National Registry. To learn more, go to **www.nccer.org** or contact us at **1.888.622.3720**. Our website has information on the latest product releases and training, as well as online versions of our *Cornerstone* newsletter and Pearson's product catalog.

Your feedback is welcome. You may email your comments to **curriculum@nccer.org,** send general comments and inquiries to **info@nccer.org**, or fill out the User Update form at the back of this module.

V.1 9/11

80201-11
ALTERNATING CURRENT AND THREE-PHASE SYSTEMS

Objectives

When you have completed this module, you will be able to do the following:

1. Describe how single-phase and three-phase alternating current is developed.
2. Calculate the peak and effective voltage or current values for an AC waveform.
3. Describe phase relationships in AC circuits.
4. Describe impedance and explain how it affects AC circuits.
5. Describe the operating principles and functions of inductors.
6. Describe the operating principles and functions of capacitors.
7. Explain the principles and functions of transformers.
8. Explain the following terms as they relate to AC circuits:
    - True power
    - Apparent power
    - Reactive power
    - Power factor

Performance Tasks

This is a knowledge-based module; there are no performance tasks.

Trade Terms

Capacitance
Frequency
Hertz (Hz)
Impedance
Inductance
Micro
Peak voltage
Reactance
Resonance
Root-mean-square (rms)
Self-inductance

Required Trainee Materials

Scientific calculator

Industry Recognized Credentials

If you're training through an NCCER-accredited sponsor you may be eligible for credentials from NCCER's Registry. The ID number for this module is 80201-11. Note that this module may have been used in other NCCER curricula and may apply to other level completions. Contact NCCER's Registry at 888.622.3720 or go to nccer.org for more information.

Contents

Topics to be presented in this module include:

Figures and Tables

1.0.0 Introduction

Alternating current (AC) reverses between positive and negative polarities and varies in amplitude with time. One complete waveform or cycle includes a complete set of variations, with two alternations in polarity. Many sources of voltage change direction with time and produce a waveform. The most common AC waveform is the sine wave.

This module covers the properties of the sine wave and other waveforms; the difference between AC and DC; the effects of **capacitance** and **inductance** on AC circuits; and the uses of transformers in AC circuits. Because of the alternating nature of AC, the components of an AC circuit react differently than those of a DC circuit. It is important to understand these differences, because they provide the basis for audio, video, and radio frequency systems.

2.0.0 Sine Wave Generation

To understand how the AC sine wave is produced, some of the basic principles learned in magnetism should be reviewed. Two principles serve the basis of all electromagnetic reactions:

- An electric current in a conductor creates a magnetic field that surrounds the conductor.
- Relative motion between a conductor and a magnetic field creates a voltage in the conductor. This occurs when at least one component of that relative motion is in a direction that is perpendicular to the direction of the field.

Figure 1 shows how these principles are applied to generate an AC waveform in a simple one-loop rotary generator. The conductor loop rotates through the magnetic field to generate the

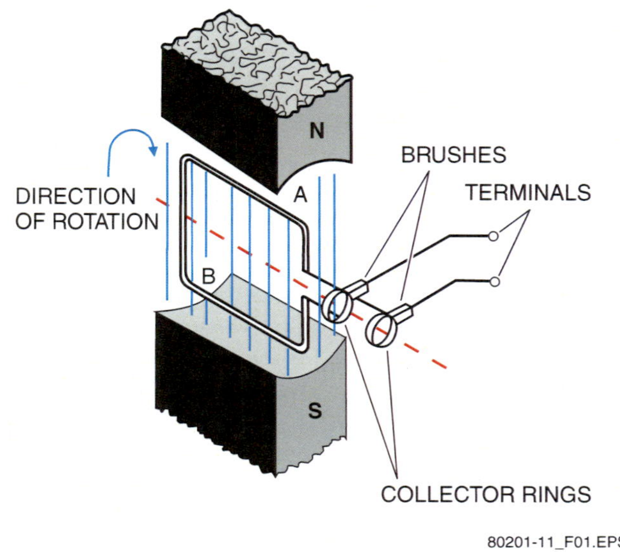

Figure 1 Conductor moving across a magnetic field.

On Site

Current Wars

After inventing the electric light bulb in 1879, Thomas Edison began work on a system for delivering electricity to homes and businesses. His system relied on direct current (DC)—electric current that always flows in one direction. However, DC transmission over long distances proved impractical. Transmitting direct current at the low voltages useful for lighting or motor operation required the use of thick, expensive copper wire. In fact, the service areas of Edison's DC generating stations were limited to about a square mile and mainly served the downtown areas of large cities. While Edison was pioneering DC distribution, electrical engineers in Europe were experimenting with alternating current (AC), which reverses direction at regular intervals.

The American businessman George Westinghouse saw the value in AC. High-voltage AC power could be distributed over longer distances using thinner, less expensive copper wires. Experts theorized that special devices (transformers) could step the voltage level up and down. Increasing the voltage level would allow it to be distributed across a wider region, while reducing the voltage level would enable the current to be used safely in homes and shops. Westinghouse hired a young electrical engineer named William Stanley, Jr., who developed the first effective transformer and demonstrated the first AC lighting system. Around the same time, the inventor Nikola Tesla filed patents for other devices run by AC.

After Westinghouse bought these patents, a full-scale industrial war known as the Current Wars erupted. At stake was whether Edison's direct current or Westinghouse's alternating current would electrify America. Edison claimed that alternating current was extremely dangerous and called for outlawing the high voltages transmitted by AC. Westinghouse countered that transformers safely reduced AC voltages before they entered buildings.

Within 10 years, the value of the alternating current system had been convincingly demonstrated. AC proved to be more practical and economical. Eventually, even Thomas Edison was forced to admit he had been wrong, and General Electric, the company he founded, began building and installing high-voltage AC transmission systems.

On Site

Why Do Power Companies Generate and Distribute AC Power Instead of DC Power?

The transformer is the key to AC power distribution. Power plants generate and distribute AC power because it permits the use of transformers, which makes power delivery more economical. Transformers used at generation plants step the AC voltage up, which decreases the current. Decreased current allows smaller-sized wires to be used for the power transmission lines. Smaller wire is less expensive and easier to support over the long distances that the power must travel from the generation plant to remotely located substations. At the substations, transformers are again used to step AC voltages back down to a level suitable for distribution to homes and businesses.

There is no such thing as a DC transformer. This means DC power would have to be transmitted at low voltages and high currents over very large wires, making the process very uneconomical. When DC is required for special applications, the AC voltage may be converted to DC voltage by using rectifiers, which make the change electrically, or by using AC motor-DC generator sets, which make the change mechanically.

induced AC voltage across its open terminals. The magnetic flux shown here is vertical.

There are several factors affecting the magnitude of voltage developed by a conductor through a magnetic field. They are the strength of the magnetic field, the length of the conductor, and the rate at which the conductor cuts directly across or at right angles to the magnetic field.

If the strength of the magnetic field and the length of the conductor making the loop are both constant, the voltage produced varies depending on the rate at which the loop cuts directly across the magnetic field.

The rate at which the conductor cuts the magnetic field depends on two things—the speed of the generator in revolutions per minute (rpm) and the angle at which the conductor travels through the field. If the generator is operated at a constant rpm, the voltage produced at any moment depends on the angle at which the conductor is cutting the field at that instant.

In *Figure 2*, the magnetic field is shown as parallel lines called lines of flux. These lines always go from the north to south poles in a generator. The motion of the conductor is shown by the large arrow.

Assume that the speed of the conductor is constant. As the angle between the flux and the conductor motion increases, the number of flux lines cut in a given time (the rate) increases. When the conductor is moving parallel to the lines of flux (angle of 0 degrees), it is not cutting any of them, and the voltage is zero.

The angle between the lines of flux and the motion of the conductor is called θ (theta). The magnitude of the voltage produced is proportional to the sine of the angle. Sine is a trigonometric function. Each angle has a sine value that never changes.

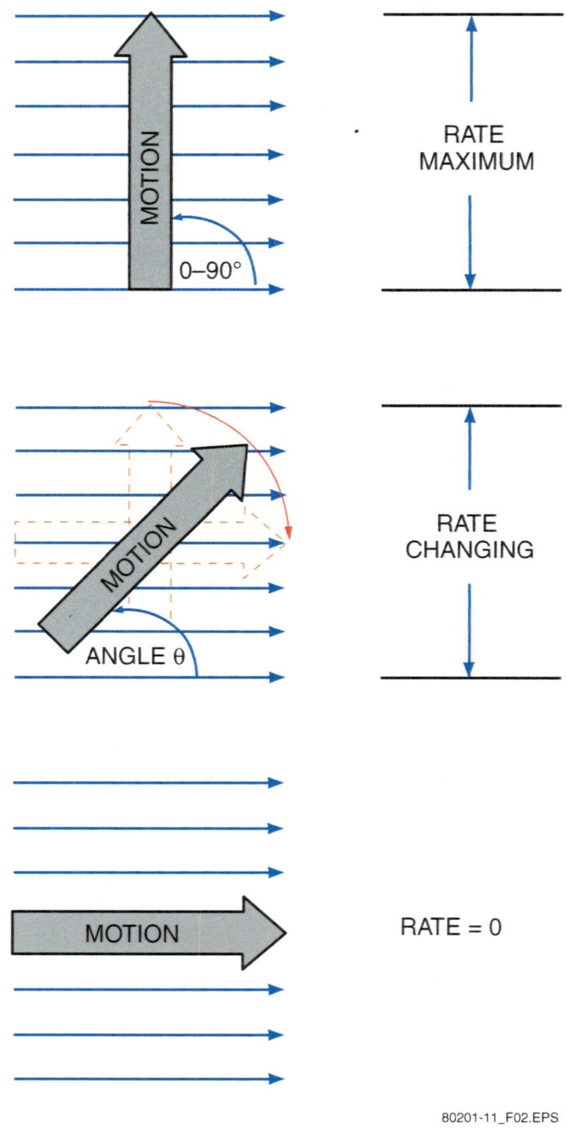

Figure 2 Angle versus rate of cutting lines of flux.

The sine of 0 degrees is 0. It increases to a maximum of 1 at 90 degrees. From 90 degrees to 180 degrees, the sine decreases back to 0. From 180 degrees to 270 degrees, the sine decreases to –1. Then from 270 degrees to 360 degrees (back to 0 degrees), the sine increases to its original 0.

Voltage is proportional to the sine of the angle. As the loop goes 360 degrees around the circle the voltage increases from 0 to its maximum at 90 degrees, back to 0 at 180 degrees, down to its maximum negative value at 270 degrees, and back up to 0 at 360 degrees, as shown in *Figure 3*.

Notice that at 180 degrees the polarity reverses. This is because the conductor has turned completely around and is now cutting the lines of flux in the opposite direction. The curve shown in *Figure 3* is called a sine wave because its shape is generated by the sine function. The value of voltage at any point along the sine wave can be calculated if the angle and the maximum obtainable voltage (E_{max}) are known. The formula used is as follows:

$$E = E_{max} \text{ sine } \theta$$

Where:

E = voltage induced
E_{max} = maximum induced voltage
θ = angle at which the voltage is induced

Using the above formula, the values of voltage anywhere along the sine wave in *Figure 4* can be calculated.

Sine values can be found using either a scientific calculator or trigonometric tables. With an E_{max} of 10 volts (V), the following values are calculated as examples:

$\theta = 0°$, sine = 0
$E = E_{max}\text{sine } \theta$
$E = (10V)(0)$
$E = 0V$

$\theta = 90°$, sine = 1.0
$E = E_{max}\text{sine } \theta$
$E = (10V)(1.0)$
$E = 10V$

$\theta = 180°$, sine = 0
$E = E_{max}\text{sine } \theta$
$E = (10V)(0)$
$E = 0V$

$\theta = 270°$, sine = –1.0
$E = E_{max}\text{sine } \theta$
$E = (10V)(-1.0)$
$E = -10V$

$\theta = 45°$, sine = 0.707
$E = E_{max}\text{sine } \theta$
$E = (10V)(0.707)$
$E = 7.07V$

$\theta = 135°$, sine = 0.707
$E = E_{max}\text{sine } \theta$
$E = (10V)(0.707)$
$E = 7.07V$

$\theta = 225°$, sine = –0.707
$E = E_{max}\text{sine } \theta$
$E = (10V)(-0.707)$
$E = -7.07V$

$\theta = 315°$, sine = –0.707
$E = E_{max}\text{sine } \theta$
$E = (10V)(-0.707)$
$E = -7.07V$

3.0.0 Sine Wave Terminology

In order to fully understand alternating current, you must understand the properties of the sine wave. The key properties are **frequency**, period, and voltage values.

3.1.0 Frequency

The frequency of a waveform is the number of times per second an identical pattern repeats itself. Each time the waveform changes from zero to a peak value and back to zero is called an al-

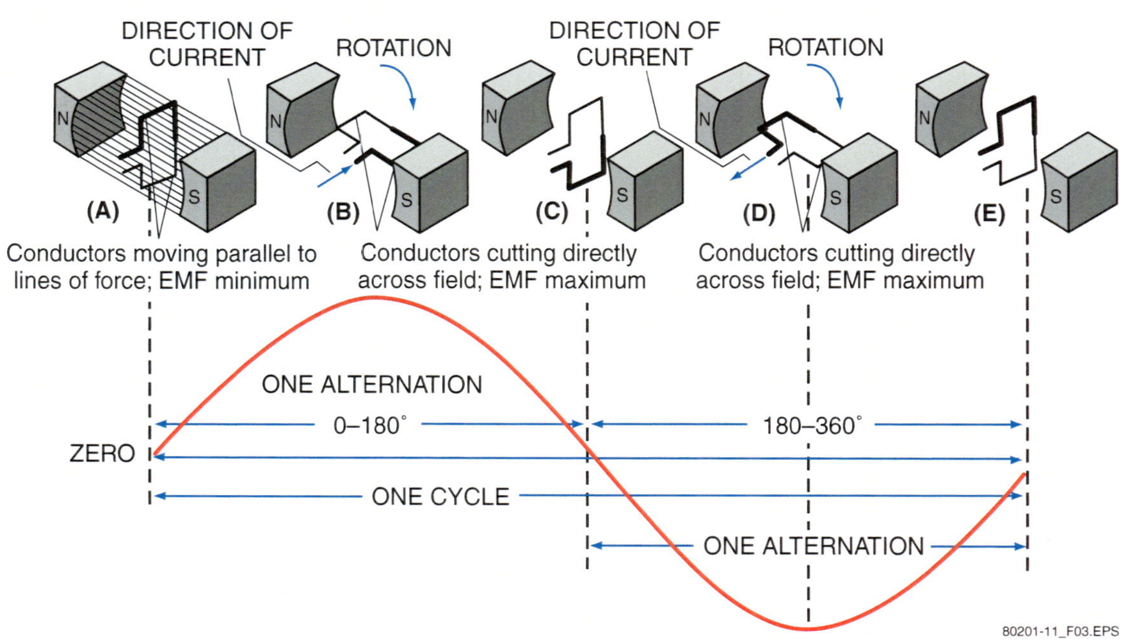

Figure 3 One cycle of alternating voltage.

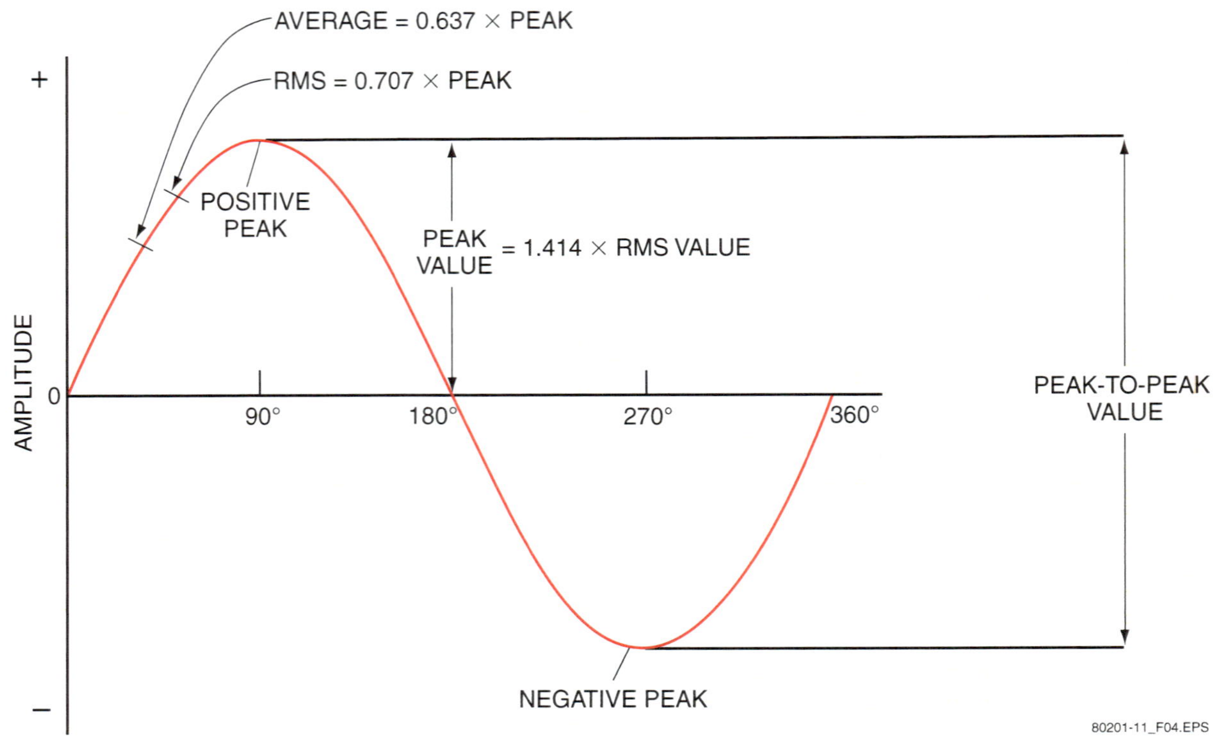

Figure 4 Amplitude values for a sine wave.

ternation. Two alternations form one cycle. The number of cycles per second is the frequency. The unit of frequency is **hertz (Hz).** One hertz equals one cycle per second (cps).

For example, determine the frequency of the waveform shown in *Figure 5*. In one-half second, the basic sine wave is repeated five times. Therefore, the frequency (f) is:

$$f = \frac{5 \text{ cycles}}{0.5 \text{ second}} = 10 \text{ cycles per second (Hz)}$$

The period of a waveform is the time (t) required to complete one cycle. The period is the inverse of frequency:

$$t = \frac{1}{f}$$

Where:

t = period (seconds)
f = frequency (Hz or cps)

For example, determine the period of the waveform in *Figure 5*. If there are five cycles in one-half second, then the frequency for one cycle is 10 cps (5 ÷ 0.5 = 10). Therefore, the period is:

$$t = \frac{1}{\text{cps}}$$

$$t = \frac{1}{10} = 0.1 \text{ second}$$

3.2.0 Peak and Effective (RMS) Voltage Values

The peak value (*Figure 4*) is the maximum value of voltage (V_M) or current (I_M). Specifying that a sine wave has a **peak voltage** of 170V applies to either the positive or the negative peak. Meters used in AC circuits read a value called the effective value. The effective value is the value of the AC current or

On Site

High-Speed Travel

Electricity travels at the speed of light, which is 186,000 miles per second. The distance from Maine to California is about 3,000 miles, so it would take an electrical current about 0.016 second (3,000/186,000) to travel that distance.

Think About It

Frequency

The frequency of the utility power generated in the United States is normally 60Hz. In some European countries and elsewhere, utility power is often generated at a frequency of 50Hz. Which of these frequencies (60Hz or 50Hz) has the shortest period?

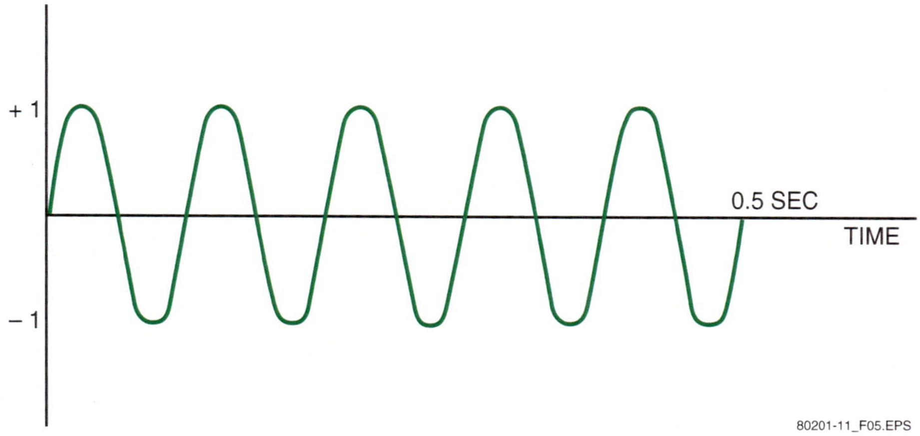

Figure 5 Frequency measurement.

voltage wave that indicates the same energy transfer as an equivalent direct current (DC) or voltage.

The direct comparison between DC and AC is in the heating effect of the two currents. Heat produced by current is a function of current amplitude only and is independent of current direction. Thus, heat is produced by both alternations of the AC wave, although the current changes direction during each alternation.

In a DC circuit, the current maintains a steady amplitude. Therefore, the heat produced is steady and is equal to I^2R. In an AC circuit, the current is continuously changing. To produce the same amount of heat from AC as from an equivalent amount of DC, the value of the AC must at times exceed the DC value.

By averaging the heating effects of all the instantaneous values during one cycle of alternating current, it is possible to find the average heat produced by the AC current during the cycle. The amount of DC required to produce that heat is equal to the effective value of the AC.

The most common method of specifying the amount of a sine wave of voltage or current is by stating its value at 45 degrees, which is 70.7 percent of the peak. This is its **root-mean-square (rms)** value. Therefore:

$$\text{Value of rms} = 0.707 \times \text{peak value}$$

For example, with a peak of 170V, the rms value is 0.707 × 170, or approximately 120V. This is the voltage of the common AC power source, which is always given in rms value.

To convert from rms voltage to peak voltage, multiply the rms value by 1.414.

Another way to view the concept of rms voltage is shown in *Figure 6*. The rms amplitude is the value assigned to an alternating voltage or current that results in the same power dissipation in a given resistance as DC voltage or current of the same amplitude. As shown, 120VAC (peak) does not produce the same light (350 lumens versus 500 lumens) as 120VDC from a 60W lamp. In order to produce the same light (500 lumens), 120V rms must be applied to the lamp. This requires that the applied sinusoidal AC waveform have a peak voltage of about 170V (170V × 0.707V = 120V).

3.3.0 Average Value

The average value is derived from all the values in a sine wave for one alternation or half cycle. The half cycle is used for the average because over

On Site

Effective Voltage

In service work, digital multimeters (DMM) like the one shown here are widely used by technicians to measure effective voltage. Unless stated otherwise, the voltages stamped on the nameplates of equipment refer to effective voltages. A true rms meter is a more sophisticated version of the DMM that reads actual rms voltage, instead of providing an estimation. Normally, a DMM estimates effective voltage because true rms is difficult to measure in an AC current that is not a perfect sine wave. The true rms meter can measure the voltage of imperfect waves.

80201-11_SA01.EPS

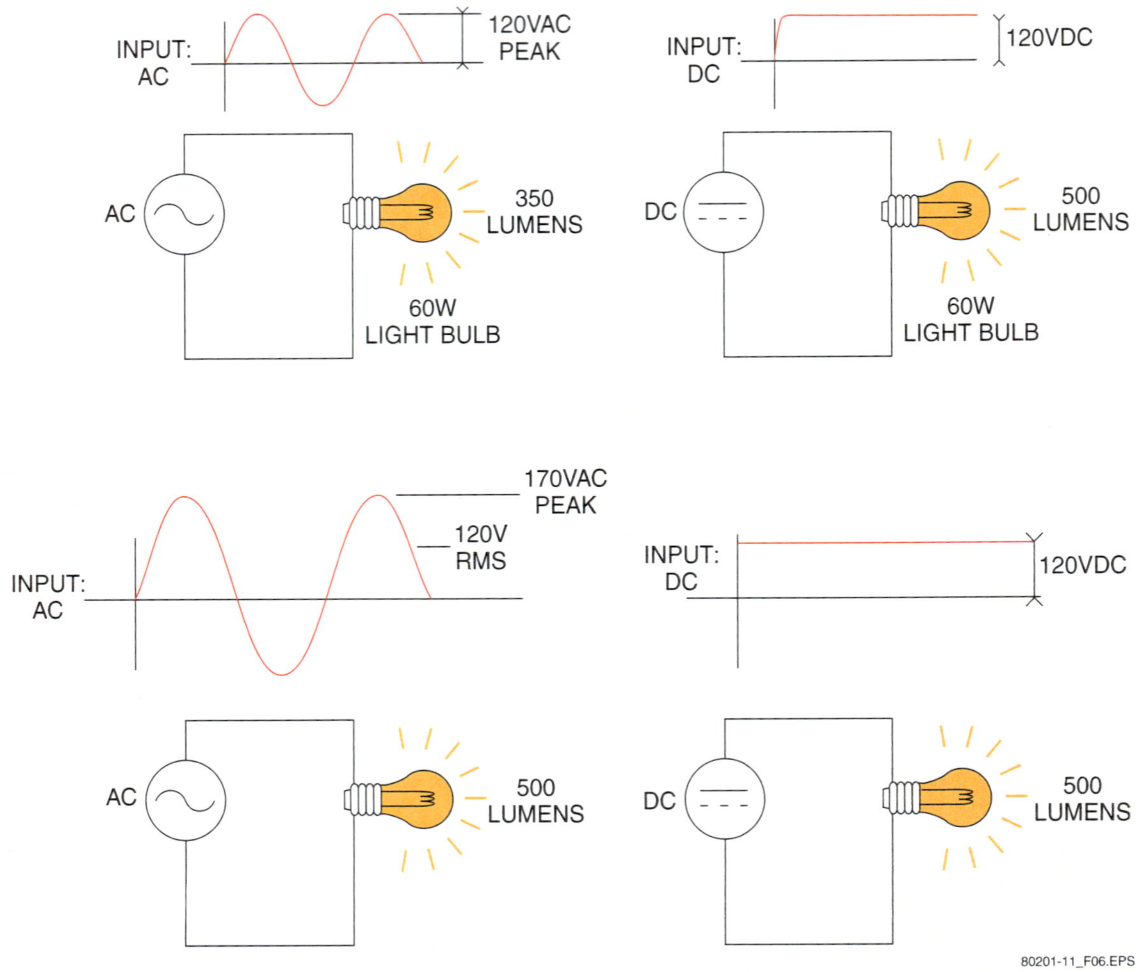

Figure 6 Root-mean-square (rms) amplitude.

a full cycle the average value is zero, which is useless for comparison purposes. If the sine values for all angles up to 180 degrees in one alternation are added and then divided by the number of values, this average equals 0.637.

Since the peak value of the sine is 1 and the average equals 0.637, the average value can be calculated as follows:

$$\text{Average value} = 0.637 \times \text{peak value}$$

For example, with a peak of 170V, the average value is 0.637 × 170V, which equals about 108V. *Figure 4* shows where the average value would fall on a sine wave.

4.0.0 AC Phase Relationships

In AC systems, phase is involved at the location of a point on a voltage or current wave with respect to the starting point of the wave or with respect to some corresponding point on the same wave. In the case of two waves of the same frequency, it is the time at which an event of one takes place with respect to a similar event of the other.

Often, the event is the starting of the waves at zero or the points at which the waves reach their maximum values. When two waves are compared in this manner, there is a phase lead or lag of one with respect to the other unless they are alternating in unison, in which case they are said to be in phase.

Suppose a generator started its cycle at 90 degrees where maximum voltage output is produced instead of starting at the point of zero output. The two output voltage waves are shown in *Figure 7*. Each is the same waveform of alternating voltage, but wave B starts at the maximum value while wave A starts at zero. The complete cycle of wave B through 360 degrees takes it back to the maximum value from which it started.

Wave A starts and finishes its cycle at zero. With respect to time, wave B is ahead of wave A in its values of generated voltage. The amount it leads in time equals one quarter revolution, which is 90 degrees. This angular difference is the phase angle between waves B and A. Wave B leads wave A by the phase angle of 90 degrees.

The 90-degree phase angle between waves B and A is maintained throughout the complete cycle and in all successive cycles as long as they both have the same frequency. At any instant in time, wave B has the value that A will have 90 degrees later. For instance, at 180 degrees, wave A is at zero, but B is already at its negative maximum value, the point where wave A will be later at 270 degrees.

To compare the phase angle between two waves, both waves must have the same frequency. Otherwise, the relative phase keeps changing. Both waves must also have sine wave variations, because this is the only kind of waveform that is measured in angular units of time. The amplitudes can be different for the two waves. The phases of two voltages, two currents, or a current with a voltage can be compared.

4.1.0 Phase Angle

To compare AC phases, it is much more convenient to use vector diagrams corresponding to the voltage waveforms, as shown in *Figure 7*. V_A and V_B represent the vector quantities corresponding to the generator voltages.

A vector is a quantity that has magnitude and direction. The length of the arrow indicates the magnitude of the alternating voltage in rms, peak, or any AC value as long as the same measure is used for all the vectors. The angle of the arrow with respect to the horizontal axis indicates the phase angle.

In *Figure 7*, the vector V_A represents the voltage wave A, with a phase angle of 0 degrees. This angle can be considered as the plane of the loop in the rotary generator where it starts with zero output voltage. The vector V_B is vertical to show the phase angle of 90 degrees for this voltage wave, corresponding to the vertical generator loop at the start of its cycle. The angle between the two vectors is the phase angle.

The symbol for a phase angle is θ (theta). In *Figure 7*, θ = 0 degrees. *Figure 8* shows the waveforms and phasor diagram of two waves that are in phase but have different amplitudes.

5.0.0 Resistance in AC Circuits

An AC circuit has an AC voltage source. Note the circular symbol with the sine wave inside it shown in *Figure 9*. It is used for any source of sine wave alternating voltage. This voltage connected across an external load resistance produces alternating current of the same waveform, frequency, and phase as the applied voltage.

According to Ohm's law, current (I) equals voltage (E) divided by resistance (R). When E is an rms value, I is also an rms value. For any instantaneous value of E during the cycle, the value of I is for the corresponding instant of time.

In an AC circuit with only resistance, the current variations are in phase with the applied voltage, as shown in *Figure 9*. This in-phase relationship between E and I means that such an AC circuit can be analyzed by the same methods used for DC circuits since there is not a phase angle to consider. Components that have only resistance include resistors, the filaments for incandescent light bulbs, and vacuum tube heaters.

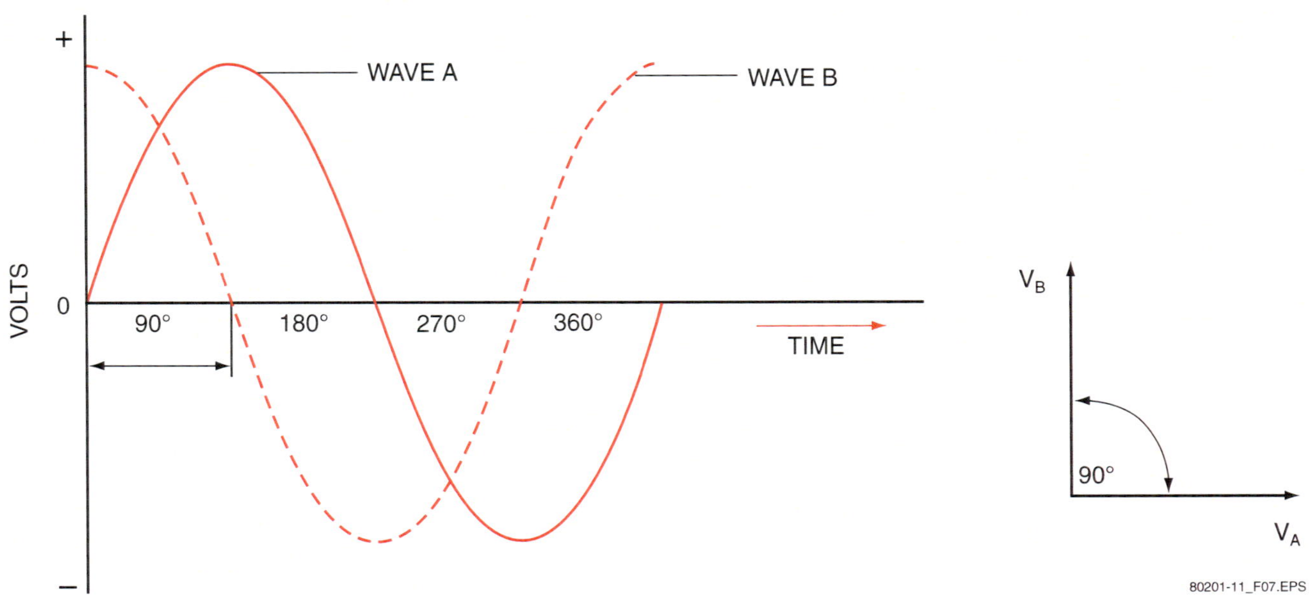

Figure 7 Voltage waveforms 90 degrees out of phase.

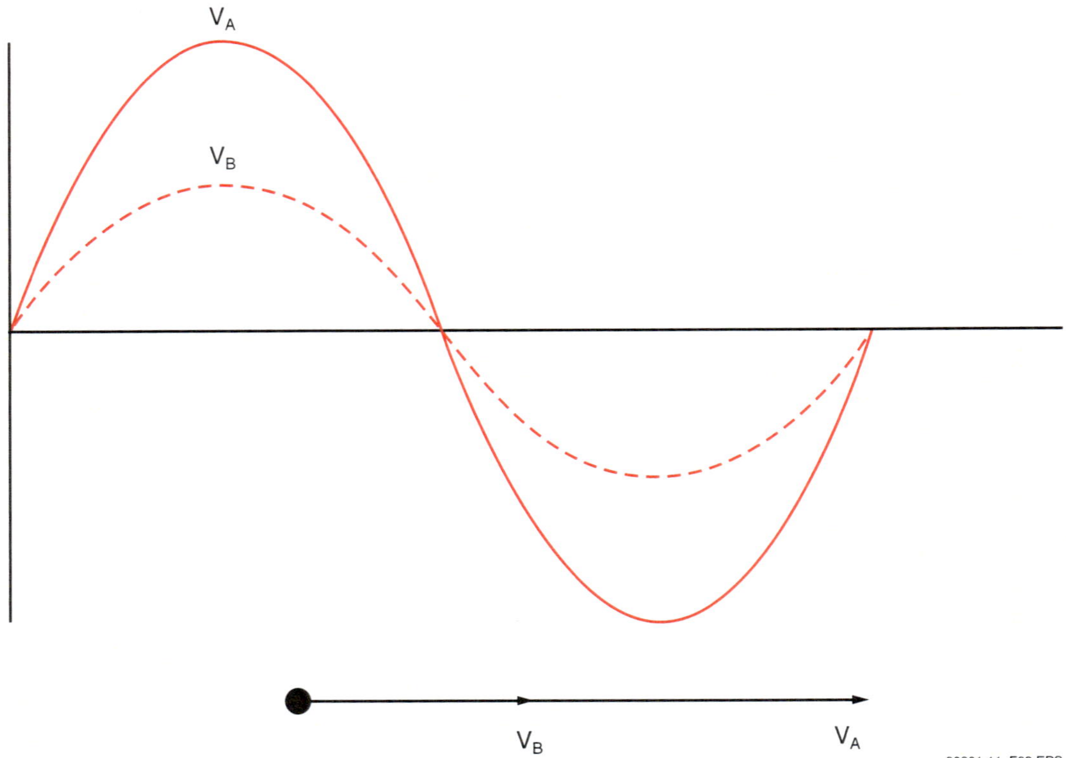

Figure 8 Waves in phase.

In purely resistive AC circuits, the voltage, current, and resistance are related by Ohm's law because the voltage and current are in phase:

$$I = \frac{E}{R}$$

Unless otherwise noted, the calculations in AC circuits are generally in rms values. For example, in *Figure 9*, the 120V applied across the 10V resistance R_L produces an rms current of 12A. This is determined as follows:

$$I = \frac{E}{R_L} = \frac{120V}{10\Omega} = 12A$$

Furthermore, the rms power (true power) dissipation is I^2R or:

$$P = (12A)^2 \times 10\Omega = 1{,}440W$$

Figure 10 shows the relationship between voltage and current in purely resistive AC circuits. The voltage and current are in phase, their cycles begin and end at the same time, and their peaks occur at the same time.

The value of the voltage shown in *Figure 10* depends on the applied voltage to the circuit. The value of the current depends on the applied voltage and the amount of resistance. If resistance is changed, it affects only the magnitude of the current.

The total resistance in any AC circuit, whether it is a series, parallel, or series-parallel circuit, is calculated using the same rules that you learned and applied to DC circuits with resistance.

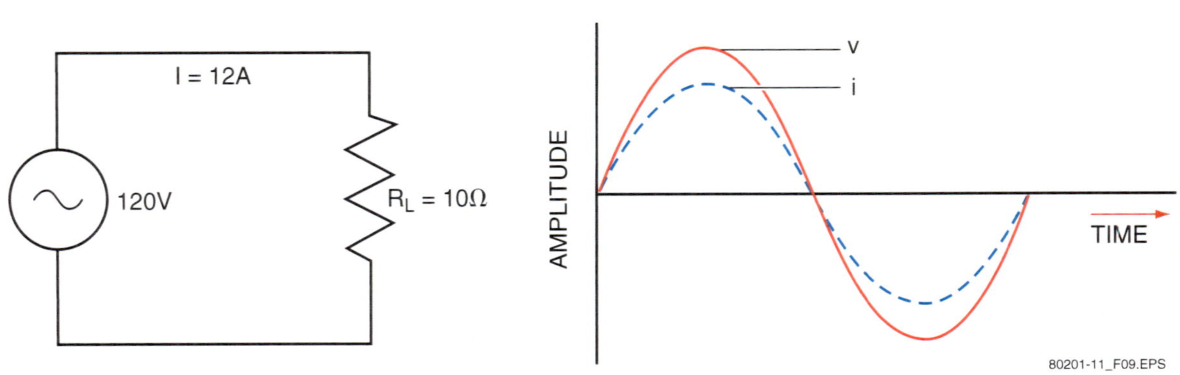

Figure 9 Resistive AC circuit.

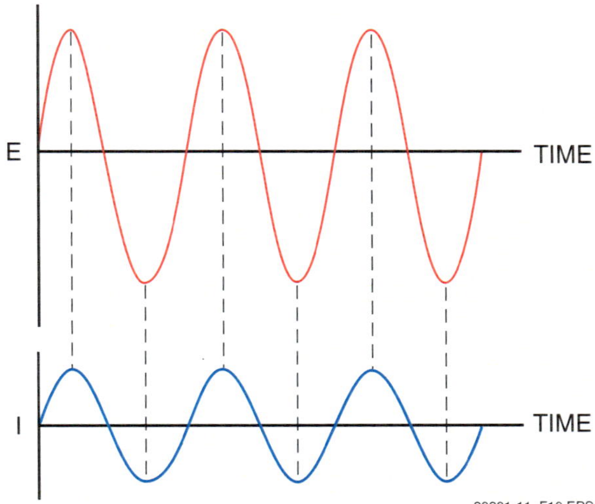

Figure 10 Voltage and current in a resistive AC circuit.

Power computation is discussed later in this module.

6.0.0 Inductance in AC Circuits

An inductor is a coil of wire wound around an iron core. As an alternating current flows through the coil, the coil becomes an electromagnet. Magnetic lines of force, called flux lines, radiate from the coil, expanding and collapsing as the AC waveform varies. The flux lines induce an opposing voltage into the coil. This induced voltage opposes the change in current flow and thereby causes the current to lag the voltage by 90 degrees. If the magnetic lines of force cut through any nearby conductor, a voltage will be induced into that conductor as well. This is the principle on which the transformer is based.

In DC circuits, a change must occur in the circuit to cause inductance. The current must change to provide motion of the flux. A steady DC of 10A cannot produce any induced voltage as long as the current value is constant. A current of 1A changing to 2A does induce voltage. Also, the faster the current changes, the higher the induced voltage becomes, because when the flux moves at a higher speed it can induce more voltage.

However, in an AC circuit the current is constantly changing and producing induced voltage. Lower frequencies of AC require more inductance to produce the same amount of induced voltage as a higher frequency current. The current can have any waveform as long as the amplitude is changing.

The ability of a conductor to induce voltage in itself when the current changes is referred to as **self-inductance** or simply inductance. The symbol for inductance is L and its unit is the henry (H). One henry is the amount of inductance that allows one volt to be induced when the current changes at the rate of one ampere per second.

6.1.0 Factors Affecting Inductance

An inductor is a coil of wire that may be wound on a core of metal or paper, or it may be self-supporting. It may consist of turns of wire placed side by side to form a layer of wire over the core or coil form. The inductance of a coil or inductor depends on its physical construction. The following factors affecting inductance are shown in *Figure 11*:

- *Number of turns* – The greater the number of turns, the greater the inductance. In addition, the spacing of the turns on a coil also affects inductance. A coil that has widely spaced turns has a lower inductance than one with the same number of more closely spaced turns. The reason for this higher inductance is that the closely wound turns produce a more concentrated magnetic field, causing the coil to show a greater inductance.
- *Coil diameter* – The inductance increases directly as the cross-sectional area of the coil increases.
- *Length of the coil* – When the length of the coil is decreased, the turn spacing is decreased, increasing the inductance of the coil.
- *Core material* – The core of the coil can be either a magnetic material (such as iron) or a non-magnetic material (such as paper or air). Coils wound on a magnetic core produce a stronger magnetic field than those with non-magnetic cores, giving them higher values of inductance. Air-core coils are used where small values of inductance are required.
- *Winding the coil in layers* – The more layers used to form a coil, the greater the effect the magnetic field has on the conductor. Layering a coil can increase the inductance.

6.2.0 Voltage and Current in an Inductive AC Circuit

The self-induced voltage across an inductance L is produced by a change in current with respect to time (DI/Dt) and can be stated as follows:

$$V_L = L\frac{\Delta I}{\Delta t}$$

Where:

Δ = change
V_L = volts
L = henrys
$\Delta I/\Delta t$ = amperes per second

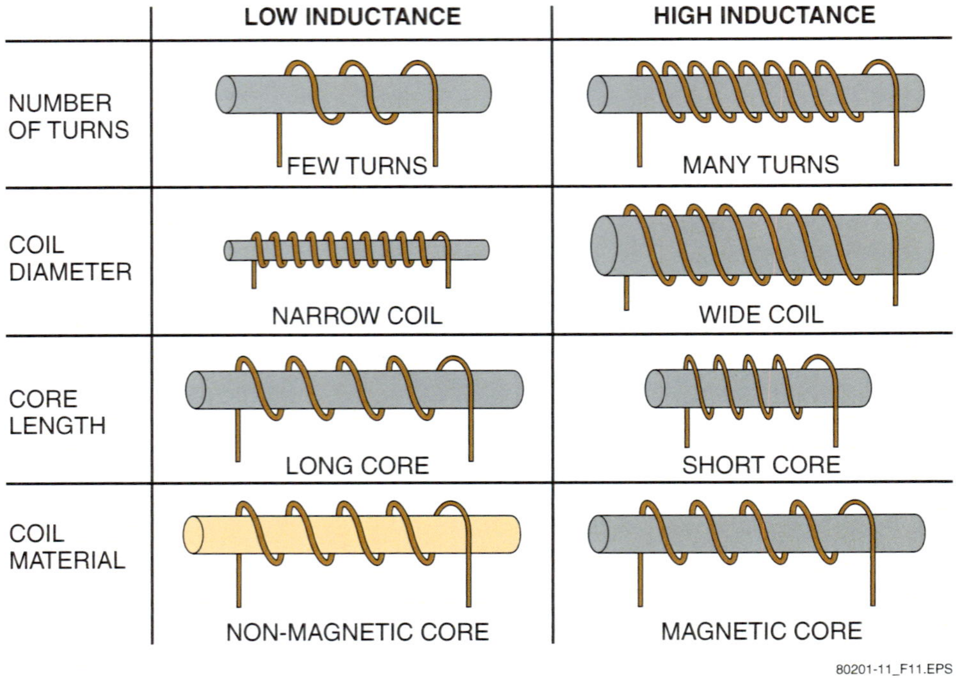

Figure 11 Factors affecting the inductance of a coil.

This gives the voltage in terms of how much magnetic flux is cut per second. When the magnetic flux associated with the current varies the same as I, this formula gives the same results for calculating induced voltage. Remember that the induced voltage across the coil is actually the result of inducing electrons to move in the conductor, so there is also an induced current.

For example, what is the self-induced voltage V_L across a 4h inductance produced by a current change of 12A per second?

$$V_L = L\frac{\Delta I}{\Delta t}$$

$$V_L = 4h\,\frac{12A}{1}$$

$$V_L = 4 \times 12$$

$$V_L = 48V$$

The current through a 200-microhenry (μh) inductor changes from 0 to 200 milliamps (mA) in 2 microseconds (μsec). The prefix **micro** means one-millionth. Determine the V_L:

$$V_L = L\frac{\Delta I}{\Delta t}$$

$$V_L = (200 \times 10^{-6})\,\frac{200 \times 10^{-3}}{2 \times 10^{-6}}$$

$$V_L = 20V$$

The induced voltage is an actual voltage that can be measured, although V_L is produced only while the current is changing. When DI/Dt is present for only a short time, V_L is in the form of a voltage pulse. With a sine wave current that is always changing, V_L is 90 degrees out of phase with I_L.

The current that flows in an inductor is induced by the changing magnetic field that surrounds the inductor. This changing magnetic field is produced by an AC voltage source that is applied to the inductor. The magnitude and polarity of the induced current depend on the field strength, direction, and rate at which the field cuts the inductor windings. The overall effect is that the current is out of phase and lags the applied voltage by 90 degrees.

At 270 degrees in *Figure 12*, the applied EMF is zero, but it is increasing in the positive direction at its greatest rate of change. Likewise, electron flow due to the applied EMF is also increasing at its greatest rate. As the electron flow increases, it produces a magnetic field that is building with it. The lines of flux cut the conductor as they move outward from it with the expanding field.

As the lines of flux cut the conductor, they induce a current into it. The induced current is at its maximum value because the lines of flux are expanding outward through the conductor at their greatest rate. The direction of the induced current opposes the force that generated it. Therefore, at 270 degrees the applied voltage is zero and is increasing to a positive value, while the current is at its maximum negative value.

At 0 degrees in *Figure 12*, the applied voltage is at its maximum positive value, but its rate of change is zero. Therefore, the field it produces

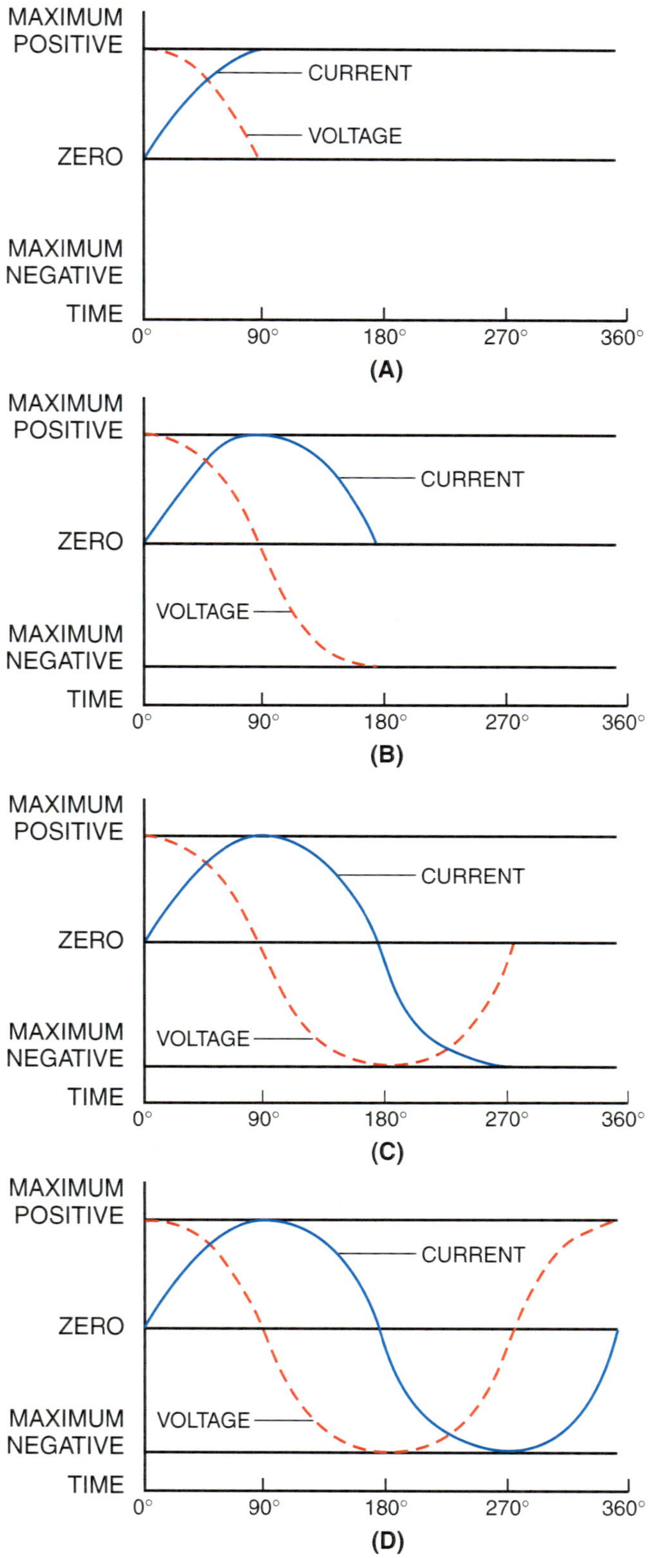

Figure 12 Inductor voltage and current relationship.

is no longer expanding and is not cutting the conductor. Because there is no relative motion between the field and conductor, no current is induced. Therefore, at 0 degrees voltage is at its maximum positive value, while current is zero.

At 90 degrees in *Figure 12*, voltage is once again zero, but this time it is decreasing toward negative at its greatest rate of change. Because the applied voltage is decreasing, the magnetic field is collapsing inward on the conductor. This has the effect of reversing the direction of motion between the field and conductor that existed at 0 degrees.

Therefore, the current will flow in a direction opposite of what it was at 0 degrees. Also, because the applied voltage is decreasing at its greatest rate, the field is collapsing at its greatest rate. This causes the flux to cut the conductor at the greatest rate, causing the induced current magnitude to be maximum. At 90 degrees, the applied voltage is zero decreasing toward negative, while the current is maximum positive.

At 180 degrees in *Figure 12*, the applied voltage is at its maximum negative value, but just as at 0 degrees, its rate of change is zero. At 180 degrees, therefore, current will be zero. This explanation shows that the voltage peaks positive first, then 90 degrees later the current peaks positive. Current thus lags the applied voltage in an inductor by 90 degrees. This can easily be remembered using this phrase: ELI the ICE man. ELI represents voltage (E), inductance (L), and current (I). In an inductor, the voltage leads the current just like the letter E leads or comes before the letter I. The word ICE will be explained in the section on capacitance.

6.3.0 Inductive Reactance

The opposing force that an inductor presents to the flow of alternating current cannot be called resistance since it is not the result of friction within a conductor. The name given to this force is inductive **reactance** because it is the reaction of the inductor to alternating current. Inductive reactance is measured in ohms and its symbol is X_L.

Remember that the induced voltage in a conductor is proportional to the rate at which magnetic lines of force cut the conductor. The greater the rate, or the higher the frequency, the greater the counter EMF (CEMF) will be. Also, the induced voltage increases with an increase in inductance; the more turns, the greater the CEMF will be. Reactance then increases with an increase of frequency and with an increase in inductance. The formula for inductive reactance is as follows:

$$X_L = 2\pi fL$$

Where:

X_L = inductive reactance in ohms
2π = a constant in which the Greek letter pi (π) represents 3.14 and 2 × pi = 6.28
f = frequency of the alternating current in hertz
L = inductance in henrys

For example, if f is equal to 60Hz and L is equal to 20h, find X_L:

$$X_L = 2\pi fL$$
$$X_L = 6.28 \times 60Hz \times 20h$$
$$X_L = 7{,}536\Omega$$

Once calculated, the value of X_L is used like resistance in a form of Ohm's law:

$$I = \frac{E}{X_L}$$

Where:

I = effective current (amps)
E = effective voltage (volts)
X_L = inductive reactance (ohms)

Unlike a resistor, there is no power dissipation in an ideal inductor. An inductor limits current, but it uses no net energy since the energy required to build up the field in the inductor is given back to the circuit when the field collapses.

7.0.0 Capacitance

Capacitance is the ability to store a charge. A capacitor is a device that stores an electric charge in a dielectric material. In storing a charge, a capacitor opposes a change in voltage. *Figure 13* shows a simple capacitor in a circuit as well as a schematic representation of two types of capacitors, and a photo of common capacitors.

Figure 14A shows a capacitor in a DC circuit. When voltage is applied, the capacitor begins to charge, as shown in *Figure 14B*. The charging continues until the potential difference across the capacitor is equal to the applied voltage. This charging current is transient or temporary since it flows only until the capacitor is charged to the applied voltage. Then there is no current in the circuit. *Figure 14C* shows this with the voltage across the capacitor equal to the battery voltage or 10V.

The capacitor can be discharged by connecting a conducting path across the dielectric. The stored charge across the dielectric provides the potential difference to produce a discharge current, as shown in *Figure 14D*. Once the capacitor is completely discharged, the voltage across it equals zero, and there is no discharge current.

In a capacitive circuit, the charge and discharge current must always be in opposite directions. Current flows in one direction to charge the capacitor and in the opposite direction when the capacitor is allowed to discharge.

Current flows in a capacitive circuit with AC voltage applied because of the capacitor charge and discharge current. There is no current through the dielectric, which is an insulator. While the capacitor is being charged by increasing applied voltage, the charging current flows in one direction to the plates. While the capacitor is discharging as the applied voltage decreases, the discharge current flows in the reverse direction. With alternating voltage applied, the capacitor alternately charges and discharges.

First, the capacitor is charged in one polarity, and then it discharges. Next, the capacitor is charged in the opposite polarity, and then it discharges again. The cycles of charge and discharge current provide alternating current in the circuit at the same frequency as the applied voltage. The

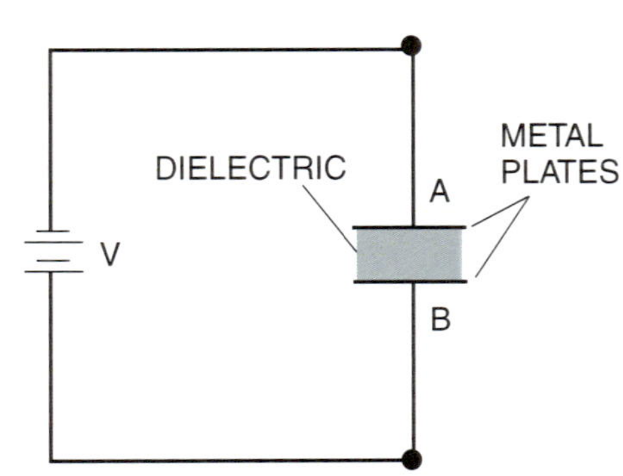

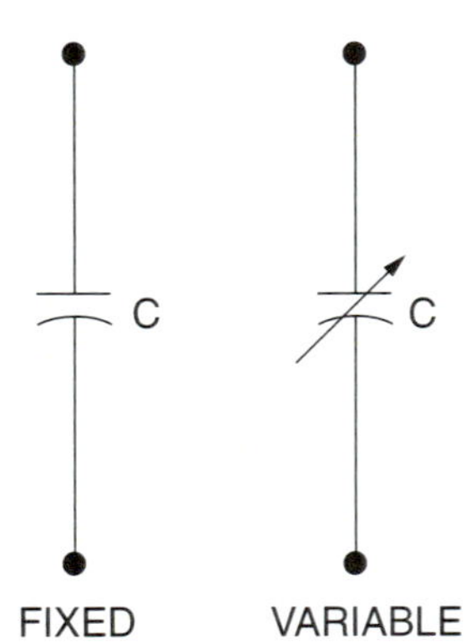

80201-11_F13.EPS

Figure 13 Capacitors.

On Site

ELI in ELI the ICE Man

Remembering the phrase ELI, as in ELI the ICE man, is an easy way to remember the phase relationships that always exist between voltage and current in an inductive circuit. An inductive circuit is a circuit where there is more inductive reactance than capacitive reactance. The L in ELI indicates inductance. The E (voltage) is stated before the I (current) in ELI, meaning that the voltage leads the current in an inductive circuit.

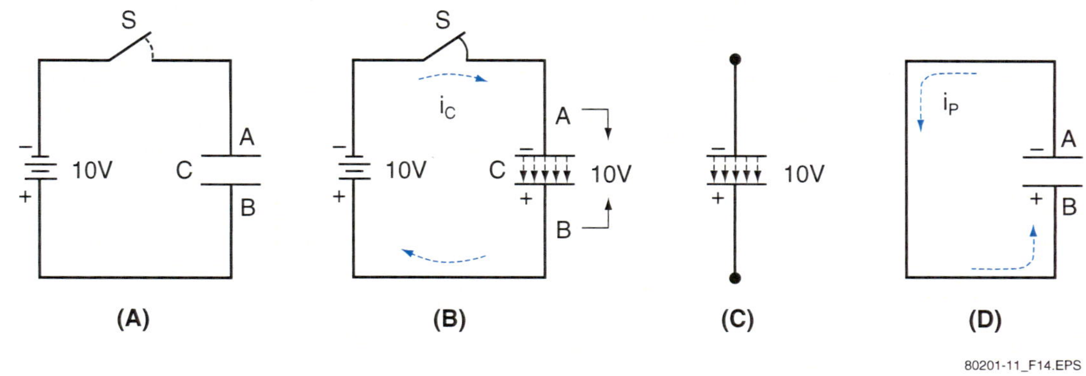

Figure 14 Charging and discharging capacitor.

amount of capacitance in the circuit determines how much current is allowed to flow.

Capacitance is measured in farads (F), where one farad is the capacitance when one coulomb is stored in the dielectric with a potential difference of one volt. Smaller values are measured in microfarads (μF). A small capacitance allows less charge and discharge current to flow than a larger capacitance. The smaller capacitor has more opposition to alternating current, because less current flows with the same applied voltage.

In summary, capacitance exhibits the following properties:

- DC is blocked by a capacitor. Once charged, no current will flow in the circuit.
- AC flows in a capacitive circuit with AC voltage applied.
- A smaller capacitance allows less current.

7.1.0 Factors Affecting Capacitance

A capacitor consists of two conductors separated by an insulating material called a dielectric. There are many types and sizes of capacitors with different dielectric materials. The capacitance is determined by three factors:

- *Area of the plates* – The initial charge displacement on a set of capacitor plates is related to the number of free electrons in each plate. Larger plates produce a greater capacitance than smaller ones. Therefore, the capacitance varies directly with the area of the plates. For example, if the area of the plates is doubled, the capacitance is doubled. If the size of the plates is reduced by 50 percent, the capacitance would also be reduced by 50 percent.
- *Distance between plates* – As two capacitor plates are brought closer together, more electrons will move away from the positively charged plate and into the negatively charged plate. This is because the mutual attraction between the opposite charges on the plates increases as the plates move closer together. This added movement of charge increases the capacitance of the capacitor. In a capacitor composed of two plates of equal area, the capacitance varies inversely with the distance between the plates. For ex-

On Site

Capacitance

The concept of capacitance, like many electrical quantities, is often hard to visualize or understand. A comparison with a balloon may help to make this concept clearer. Electrical capacitance has a charging effect similar to blowing up a balloon and holding it closed. The expansion capacity of the balloon can be changed by changing the thickness of the balloon walls. A balloon with thick walls will expand less (have less capacity) than one with thin walls. This is like a small 10μF capacitor that has less capacity and will charge less than a larger 100μF capacitor.

ample, if the distance between the plates is decreased by one-half, the capacitance is doubled. If the distance between the plates is doubled, the capacitance would be one-half as great.

- *Dielectric permittivity* – Another factor that determines the value of capacitance is the permittivity of the dielectric. The dielectric is the material between the capacitor plates in which the electric field appears. Relative permittivity expresses the ratio of the electric field strength in a dielectric to that in a vacuum. Permittivity has nothing to do with the dielectric strength of the medium or the breakdown voltage. An insulating material that withstands a higher applied voltage than some other substance does not always have a higher dielectric permittivity. Many insulating materials have a greater dielectric permittivity than air. For a given applied voltage, a greater attraction exists between the opposite charges on the capacitor plates, and an electric field can be set up more easily than when the dielectric is air. The capacitance of the capacitor is increased when the permittivity of the dielectric is increased if all the other factors remain unchanged.

7.2.0 Calculating Equivalent Capacitance

Connecting capacitors in parallel is equivalent to adding the plate areas. Therefore, the total capacitance is the sum of the individual capacitances, as shown in *Figure 15*.

A 10μF capacitor in parallel with a 5μF capacitor, for example, provides a 15μF capacitance for the parallel combination. The voltage is the same across the parallel capacitors. Note that adding parallel capacitance is opposite to the case of inductances in parallel and resistances in parallel.

Connecting capacitances in series is equivalent to increasing the thickness of the dielectric. Therefore, the combined capacitance is less than the smallest individual value. The combined equivalent capacitance is calculated by the reciprocal formula, as shown in *Figure 16*.

Capacitors connected in series are combined like resistors in parallel. Any of the shortcut calculations for the reciprocal formula apply. For example, the combined capacitance of two equal capacitances of 10μF in series is 5μF. Capacitors are used in series to provide a higher voltage breakdown rating for the combination. For instance, each of three equal capacitances in series has one-third the applied voltage.

In series, the voltage across each capacitor is inversely proportional to its capacitance. The smaller capacitance has the larger proportion of the applied voltage. The reason is that the series capacitances all have the same charge because they are in one current path. With equal charge, a smaller capacitance has a greater potential difference.

Think About It

Capacitor Substitution

Suppose you had a motor with a bad 30μF starting capacitor, and no 30μF direct replacement capacitor was available. As a temporary measure, you are authorized to substitute two equal-value capacitors in its place. What size capacitors (μF) should be used if you are connecting them in parallel?

$$C_T = C_1 + C_2$$

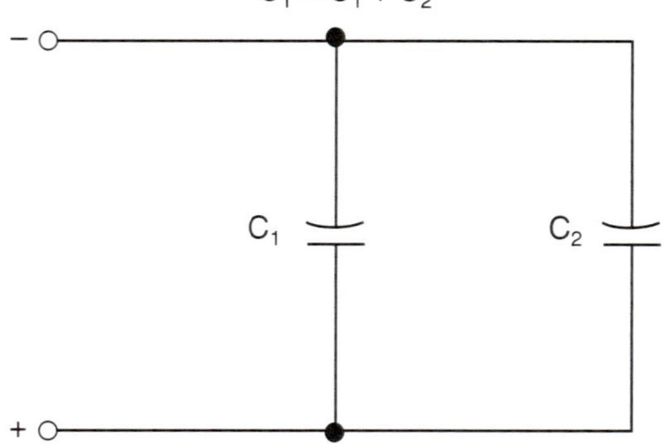

Figure 15 Capacitors in parallel.

$$C_T = \frac{1}{\frac{1}{C_1} + \frac{1}{C_2}}$$

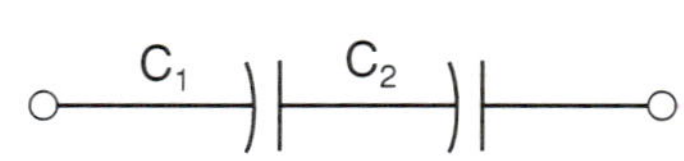

Figure 16 Capacitors in series.

7.3.0 Capacitor Specifications

In addition to its capacitance rating, the capacitor is rated by its operating voltage and leakage resistance. These factors must be considered in the selection of a capacitor.

7.3.1 Voltage Rating

This rating specifies the maximum potential difference that can be applied across the plates without puncturing the dielectric. Usually, the voltage rating is for temperatures up to about 60°C. High temperatures result in a lower voltage rating. Voltage ratings for general-purpose paper, mica, and ceramic capacitors are typically 200V to 500V. Ceramic capacitors with ratings of 1 to 5kV are also available.

Electrolytic capacitors are commonly used in 25V, 150V, and 450V ratings. In addition, 6V and 10V electrolytic capacitors are often used in transistor circuits. For applications where a lower voltage rating is permissible, more capacitance can be obtained in a smaller physical size.

The potential difference across the capacitor depends on the applied voltage and is not necessarily equal to the voltage rating. A voltage rating higher than the potential difference applied across the capacitor provides a safety factor for long life in service. With electrolytic capacitors, however, the actual capacitor voltage should be close to the rated voltage to produce the oxide film that provides the specified capacitance.

The voltage ratings are for applied DC voltage. The breakdown rating is lower for AC voltage because of the internal heat produced by continuous charge and discharge.

7.3.2 Leak Resistance

Consider a capacitor charged by a DC voltage source. After the charging voltage is removed, a perfect capacitor would keep its charge indefinitely. After a long period of time, however, the charge will be neutralized by a small leakage current through the dielectric and across the insulated case between terminals, because there is no perfect insulator. For paper, ceramic, and mica capacitors, the leakage current is very slight, or inversely, the leakage resistance is very high. For these types of capacitors, leakage resistance is 100MV or more. However, electrolytic capacitors may have a leakage resistance of 0.5MV or less.

WARNING!

The residual charge on a capacitor can be dangerous even after the power is turned off. The charge must be bled off before the capacitor terminals can be touched.

7.4.0 Voltage and Current in a Capacitive AC Circuit

In a capacitive circuit driven by an AC voltage source, the voltage is continuously changing. Thus, the charge on the capacitor is also continuously changing. The four parts of *Figure 17* show the variation of the alternating voltage and current in a capacitive circuit for each quarter of one cycle.

The solid line represents the voltage across the capacitor, and the dotted line represents the current. The line running through the center is the zero or reference point for both the voltage and the current. The bottom line marks off the time of the cycle in terms of electric degrees. Assume that the AC voltage has been acting on the capacitor for some time before the time represented by the starting point of the sine wave.

At the beginning of the first quarter-cycle (0 to 90 degrees), the voltage has just passed through zero and is increasing in the positive direction. Since the zero point is the steepest part of the sine wave, the voltage is changing at its greatest rate.

The charge on a capacitor varies directly with the voltage; therefore, the charge on the capacitor is also changing at its greatest rate at the beginning of the first quarter-cycle. In other words, the greatest number of electrons are moving off one plate and onto the other plate. Thus, the capacitor current is at its maximum value.

As the voltage proceeds toward maximum at 90 degrees, its rate of change becomes lower and lower, making the current decrease toward zero. At 90 degrees, the voltage across the capacitor is maximum, the capacitor is fully charged, and there is no further movement of electrons from plate to plate. That is why the current at 90 degrees is zero.

At the end of the first quarter-cycle, the alternating voltage stops increasing in the positive direction and starts to decrease. It is still a positive voltage; but to the capacitor, the decrease in voltage means that the plate that has an excess of electrons, and must lose some of them. The current flow must reverse its direction. The second part of the figure shows the current curve to be below the zero line (negative current direction) during the second quarter-cycle (90 to 180 degrees).

At 180 degrees, the voltage has dropped to zero. This means that, for a brief instant, the electrons are equally distributed between the two plates; the current is maximum because the rate of change of voltage is maximum.

Just after 180 degrees, the voltage has reversed polarity and starts building to its maximum negative peak, which is reached at the end of the third quarter-cycle (180 to 270 degrees). During the third quarter-cycle, the rate of voltage change gradually decreases as the charge builds to a maximum at 270 degrees. At this point, the capacitor is fully charged and carries the full impressed voltage. Because the capacitor is fully charged, there is no further exchange of electrons and the current flow is zero at this point. The conditions are exactly the same as at the end of the first quarter-cycle (90 degrees), but the polarity is reversed.

Just after 270 degrees, the impressed voltage once again starts to decrease, and the capaci-

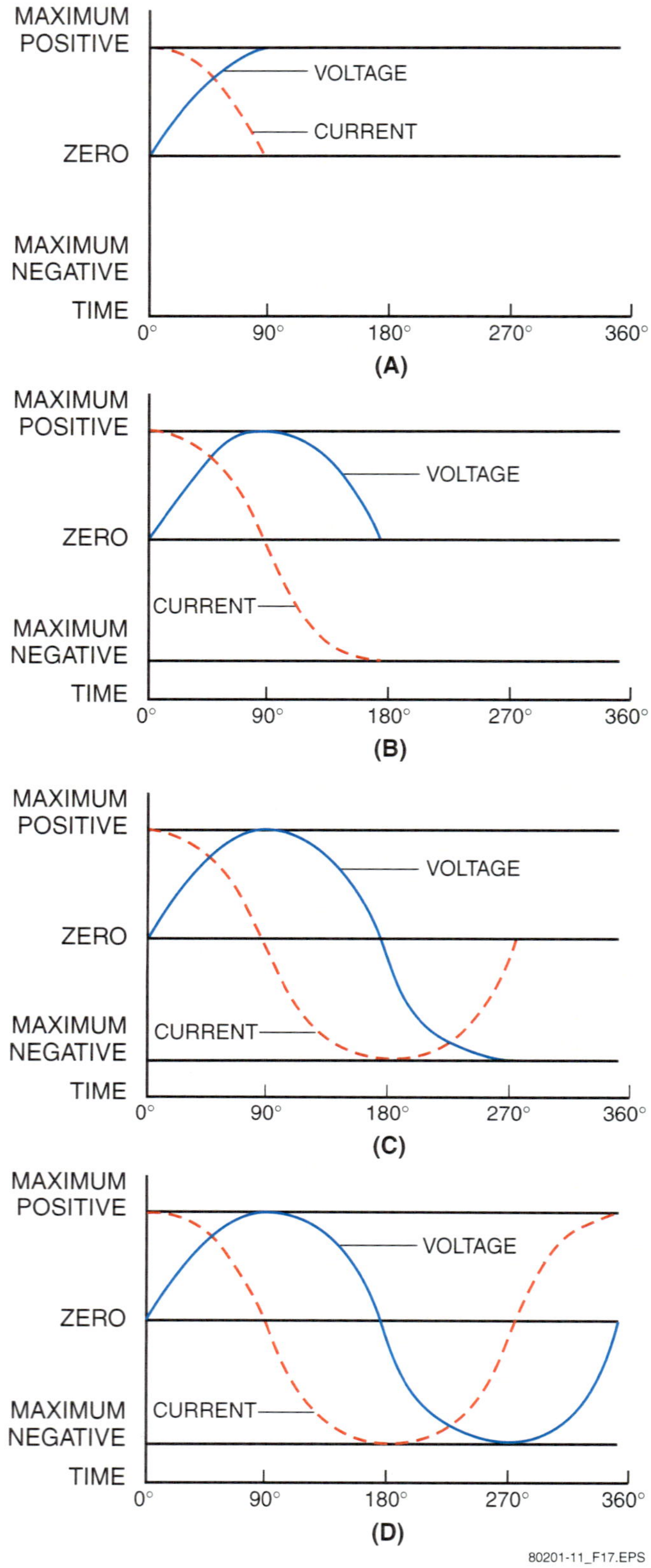

Figure 17 Voltage and current in a capacitive AC circuit.

tor must lose electrons from the negative plate. It must discharge, starting at a minimum rate of flow and rising to a maximum. This discharging action continues through the last quarter-cycle (270 to 360 degrees) until the impressed voltage has reached zero. The beginning of the entire cycle is 360 degrees, and everything starts over again.

In *Figure 17,* note that the current always arrives at a certain point in the cycle 90 degrees ahead of the voltage because of the charging and discharging action. This voltage-current phase relationship in a capacitive circuit is exactly opposite to that in an inductive circuit. The current through a capacitor leads the voltage across the capacitor by 90 degrees. A convenient way to remember this is the phrase ELI the ICE man (ELI refers to inductors, as previously explained). ICE pertains to capacitors as follows:

I = current
C = capacitor
E = voltage

In capacitors (C), current (I) leads voltage (E) by 90 degrees.

It is important to realize that the current and voltage are both going through their individual cycles at the same time during the period the AC voltage is impressed. The current does not go through part of its cycle (charging or discharging) and then stop and wait for the voltage to catch up. The amplitude and polarity of the voltage and the amplitude and direction of the current are continually changing.

Their positions, with respect to each other and to the zero line at any electrical instant or any degree between 0 and 360 degrees, can be seen by reading upward from the time-degree line. The current swing from the positive peak at 0 degrees to the negative peak at 180 degrees is not a measure of the number of electrons or the charge on the plates. It is a picture of the direction and strength of the current in relation to the polarity and strength of the voltage appearing across the plates.

7.5.0 Capacitive Reactance

Capacitors offer a very real opposition to current flow. This opposition arises from the fact that, at a given voltage and frequency, the number of electrons that go back and forth from plate to plate is limited by the storage ability or the capacitance of the capacitor. As the capacitance is increased, a greater number of electrons changes plates every cycle. Since current is a measure of the number of electrons passing a given point in a given time, the current is increased.

Increasing the frequency also decreases the opposition offered by a capacitor. This occurs because the number of electrons that the capacitor is capable of handling at a given voltage changes plates more often. As a result, more electrons pass a given point in a given time (greater current flow). The opposition that a capacitor offers to alternating current is therefore inversely proportional to frequency and capacitance. This opposition is called capacitive reactance. Capacitive reactance decreases with increasing frequency or, for a given frequency, the capacitive reactance decreases with increasing capacitance. The symbol for capacitive reactance is X_C. The formula is:

$$X_c = \frac{1}{2\pi fC}$$

Where:

X_C = capacitive reactance in ohms
f = frequency in hertz
C = capacitance in farads
$2\pi = 6.28\ (2 \times 3.14)$

For example, what is the capacitive reactance of a 0.05μF capacitor in a circuit whose frequency is 1 megahertz?

$$X_c = \frac{1}{2\pi fC}$$

$$X_c = \frac{1}{6.28(10^6 \text{ hertz})(5 \times 10^{-8} \text{ farads})}$$

$$X_c = \frac{1}{6.28(5 \times 10^{-2})}$$

$$X_c = \frac{1}{31.4 \times 10^{-2}}$$

$$X_c = \frac{1}{0.314} = 3.18 \text{ ohms}$$

The capacitive reactance of a 0.05μF capacitor operated at a frequency of 1 megahertz is 3.18 ohms. Suppose this same capacitor is operated at a lower frequency of 1,500 hertz instead of 1 megahertz. What is the capacitive reactance now? Substituting where $1{,}500 = 1.5 \times 10^3$ hertz:

$$X_c = \frac{1}{2\pi fC}$$

$$X_c = \frac{1}{6.28(1.5 \times 10^3 \text{ hertz})(5 \times 10^{-8} \text{ farads})}$$

$$X_c = \frac{1}{6.28(7.5 \times 10^{-5})}$$

$$X_c = \frac{1}{47.1 \times 10^{-5}}$$

$$X_c = \frac{1}{0.000471} = 2{,}123 \text{ ohms}$$

Note a very interesting point from these two examples. As frequency is decreased from 1 megahertz to 1,500 hertz, the capacitive reactance increases from 3.18 ohms to 2,123 ohms. Capacitive reactance increases as the frequency decreases.

8.0.0 RL, RC, LC, AND RLC CIRCUITS

AC circuits often contain inductors, capacitors, and/or resistors connected in series or parallel combinations. When this is done, it is important to determine the phase relationship between the applied voltage and the current in the circuit. The simplest method of combining factors that have different phase relationships is vector addition. Each quantity is represented as a vector, and the resultant vector and phase angle are then calculated.

In purely resistive circuits, the voltage and current are in phase. In inductive circuits, the voltage leads the current by 90 degrees. In capacitive circuits, the current leads the voltage by 90 degrees. *Figure 18* shows the phase relationships of these components used in AC circuits. Recall that these properties are summarized by the phrase ELI the ICE man.

ELI = E Leads I (inductive)
ICE = I Capacitive (leads) E

The **impedance** Z of a circuit is defined as the total opposition to current flow. The magnitude of the impedance Z is given by the following equation in a series circuit:

$$Z = \sqrt{R^2 + X^2}$$

Where:

Z = impedance (ohms)
R = resistance (ohms)
X = net reactance (ohms)

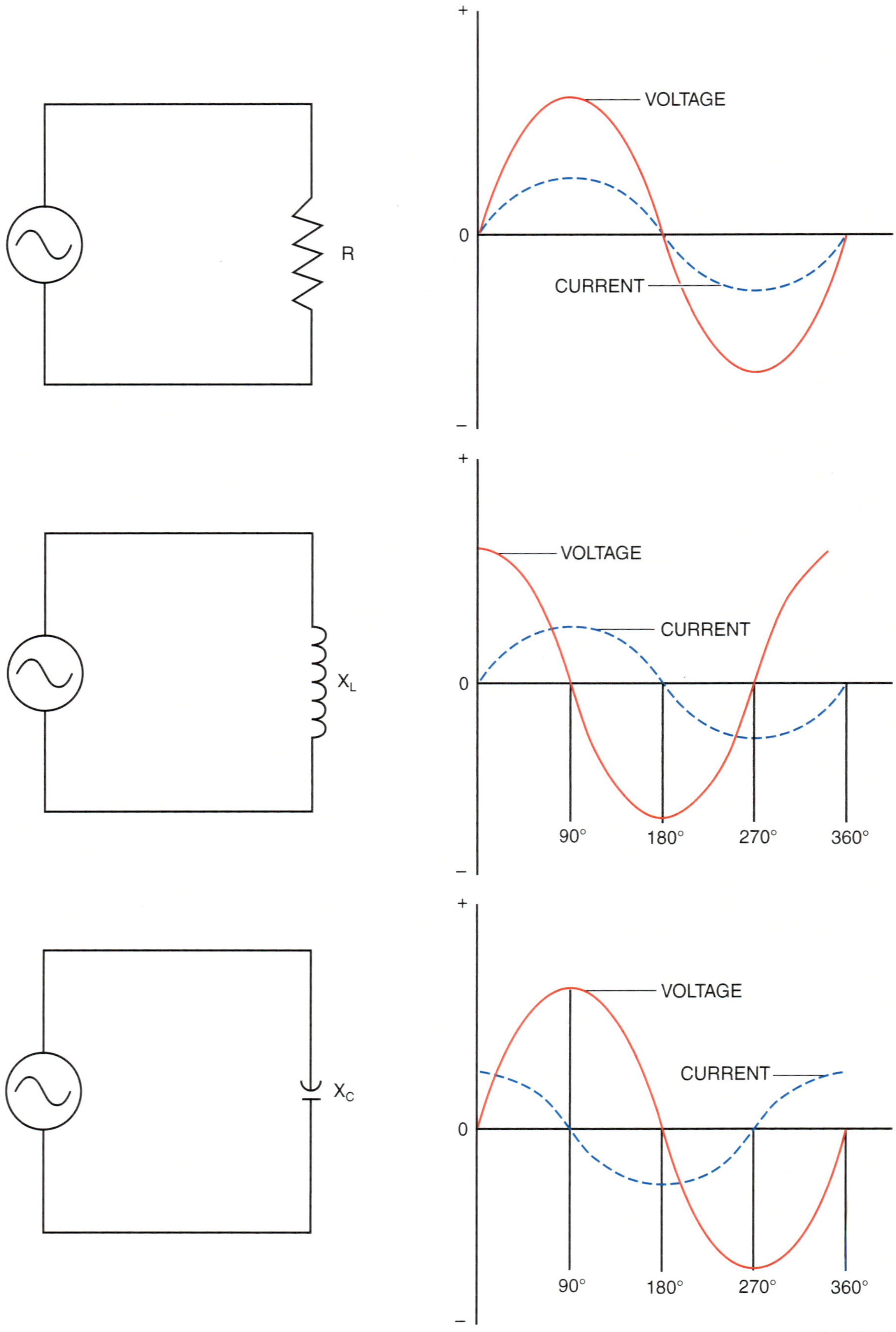

Figure 18 Summary of AC circuit phase relationships.

Think About It

Frequency and Capacitive Reactance

A variable capacitor is used in the tuner of an AM radio to tune the radio to the desired station. Will its capacitive reactance value be higher or lower when it is tuned to the low end of the frequency band (550kHz) than it would be when tuned to the high end of the band (1,440kHz)?

The current through a resistance is always in phase with the voltage applied to it; thus resistance is shown along the 0-degree axis. The voltage across an inductor leads the current by 90 degrees; thus inductive reactance is shown along the 90-degree axis. The voltage across a capacitor lags the current by 90 degrees; thus capacitive reactance is shown along the 90-degree axis. The net reactance is the algebraic difference between the inductive reactance and the capacitive reactance:

$$X = \text{net reactance (ohms)}$$
$$X_L = \text{inductive reactance (ohms)}$$
$$X_C = \text{capacitive reactance (ohms)}$$

The impedance Z is the vector sum of the resistance R and the net reactance X. The angle, called the phase angle, gives the phase relationship between the applied voltage and current.

8.1.0 Resonance

Every circuit containing inductors and capacitors has a frequency at which **resonance** occurs. This is the frequency at which the X_L and X_C are equal. In a series circuit, the impedances oppose each other and therefore cancel each other out at the resonant frequency. Because of this cancelling effect, the voltage output of the circuit is maximized. A parallel resonant circuit works the opposite: impedance is maximum at the resonant frequency, and therefore the voltage is minimized. Practical applications of this principle can be seen in filters. In a series LC filter, signals at the resonant frequency or a band of frequencies pass readily, while others are rejected. In a parallel LC filter, signals at the resonant frequency are rejected while others are passed.

9.0.0 Power In AC Circuits

In DC circuits, the power consumed is the sum of all the I^2R heating in the resistors. It is also equal to the power produced by the source, which is the product of the source voltage and current. In AC circuits containing only resistors, the same relationship also holds true.

9.1.0 True Power

The power consumed by resistance is called true power and is measured in watts. True power is the product of the resistor current squared and the resistance:

$$P_T = I^2R$$

This formula applies because current and voltage have the same phase across a resistance.

To find the corresponding value of power as a product of voltage and current, this product must be multiplied by the cosine of the phase angle θ:

$$P_T = I^2R \text{ or } P_T = EI \cos \theta$$

Where E and I are in rms values to calculate the true power in watts, multiplying I by the cosine of the phase angle provides the resistive component for true power equal to I^2R.

For example, a series RL circuit has 2A through a 100V resistor in series with the X_L of 173V. Therefore:

$$P_T = I^2R$$
$$P_T = 4 \times 100$$
$$P_T = 400W$$

Furthermore, in this circuit the phase angle is 60 degrees with a cosine of 0.5. The applied voltage is 400V. Therefore:

$$P_T = EI \times \cos \theta$$
$$P_T = 400 \times 2 \times 0.5$$
$$P_T = 400W$$

In both cases, the true power is the same (400W) because this is the amount of power supplied by the generator and dissipated in the resistance. Either formula can be used for calculating the true power.

9.2.0 Apparent Power

In ideal AC circuits containing resistors, capacitors, and inductors, the only mechanism for power consumption is $I^2_{eff}R$ heating in the resistors. Inductors and capacitors consume no power. The only function of inductors and capacitors is to store and release energy. However, because of the phase shifts that are introduced by these elements, the power consumed by the resistors is not equal to the product of the source voltage and current. The product of the source voltage and current is called apparent power and has units of volt-amperes (VA).

The apparent power is the product of the source voltage and the total current. Therefore, apparent power is actual power delivered by the source. The formula for apparent power is:

$$P_A = (E_A)(I)$$

Figure 19 shows a series RL circuit and its associated vector diagram.

This formula is used to calculate the apparent power and compare it to the circuit's true power:

$$P_A = (E_A)(I)$$
$$P_A = (400V)(2A)$$
$$P_A = 800VA$$
$$P_T = EI \times \cos\theta$$
$$\theta = \frac{R}{X_L} = \frac{173}{100} = 60°$$

$$P_T = (400V)(2A)(\cos 60°)$$
$$P_T = (400V)(2A)(0.5)$$
$$P_T = 400W$$

Note that the apparent power formula is the product of EI alone without considering the cosine of the phase angle.

9.3.0 Reactive Power

Reactive power is that portion of the apparent power that is caused by inductors and capacitors in the circuit. Inductance and capacitance are always present in real AC circuits. No work is performed by reactive power; the power is stored in the inductors and capacitors, then returned to the circuit. Therefore, reactive power is always 90 degrees out of phase with true power. The units for reactive power are volt-amperes-reactive (VARs).

In general, for any phase angle θ between E and I, multiplying EI by sine θ gives the vertical component at 90 degrees for the value of the VARs. In *Figure 19*, the value of sine 60 degrees is $800 \times 0.866 = 692.8$ VARs.

Note that the factor sine θ for the volt-amperes-reactive (VARs) gives the vertical or reactive component of the apparent power EI. However, multiplying EI by cosine θ as the power factor gives the horizontal or resistive component for the real power.

9.4.0 Power Factor

Because it indicates the resistive component, cosine θ is the power factor (pf) of the circuit, converting the EI product to real power. For series circuits, use the following formula:

$$pf = \cos\theta = \frac{R}{Z}$$

For parallel circuits, use this formula:

$$pf = \cos\theta = \frac{I_R}{I_T}$$

In *Figure 19* as an example of a series circuit, R and Z are used for the calculations:

$$pf = \cos\theta = \frac{R}{Z}$$

$$pf = \frac{100\Omega}{200\Omega} = 0.5$$

The power factor is not an angular measure but a numerical ratio with a value between 0 and 1, equal to the cosine of the phase angle. With all re-

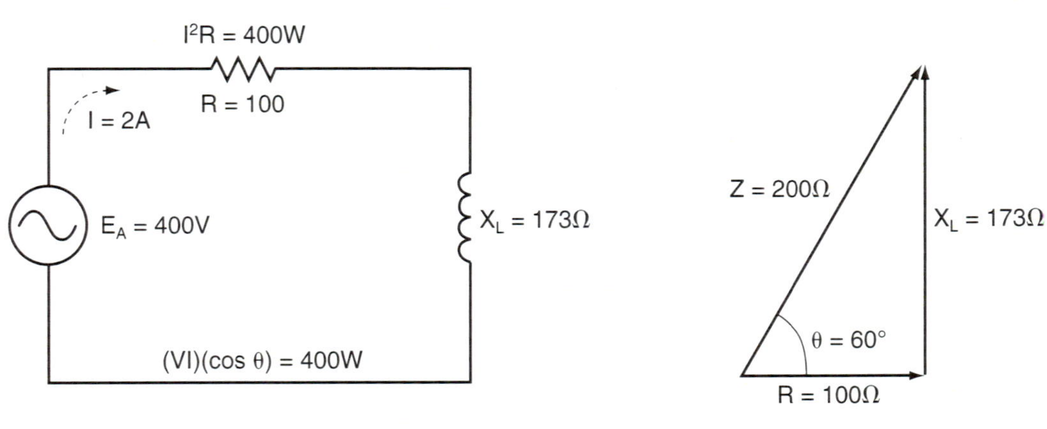

Figure 19 Power calculations in an AC circuit.

sistance and zero reactance, R and Z are the same for a series circuit of I_R and I_T and are the same for a parallel circuit. The ratio is 1. Therefore, unity power factor means a resistive circuit. At the opposite extreme, all reactance with zero resistance makes the power factor zero, meaning that the circuit is all reactive.

The power factor gives the relationship between apparent power and true power. The power factor can thus be defined as the ratio of true power to apparent power:

$$pf = \frac{P_T}{P_A}$$

For example, calculate the power factor of the circuit shown in *Figure 20*.

The true power is the product of the resistor current squared and the resistance:

$$P_T = I^2R$$
$$P_T = 10A^2 \times 10\Omega$$
$$P_T = 1{,}000W$$

The apparent power is the product of the source voltage and total current:

$$P_A = (I_T)(E)$$
$$P_A = 10.2A \times 100V$$
$$P_A = 1{,}020VA$$

Calculating total current:

$$I_T = \sqrt{I_R^2 + (I_C - I_L)^2} = \sqrt{10A^2 + (4A - 2A)^2}$$
$$I_T = 10.2A$$

The power factor is the ratio of true power to apparent power:

$$pf = \frac{P_T}{P_A}$$

$$pf = \frac{1{,}000}{1{,}020}$$

$$pf = 0.98$$

As illustrated in the previous example, the power factor is determined by the system load. If the load contained only resistance, the apparent power would equal the true power and the power factor would be at its maximum value of one. Purely resistive circuits have a power factor of unity or one. If the load is more inductive than capacitive, the apparent power will lag the true power and the power factor lags. If the load is more capacitive than inductive, the apparent power leads the true power and the power factor leads. If there is any reactive load on the system, the apparent power is greater than the true power and the power factor is less than one.

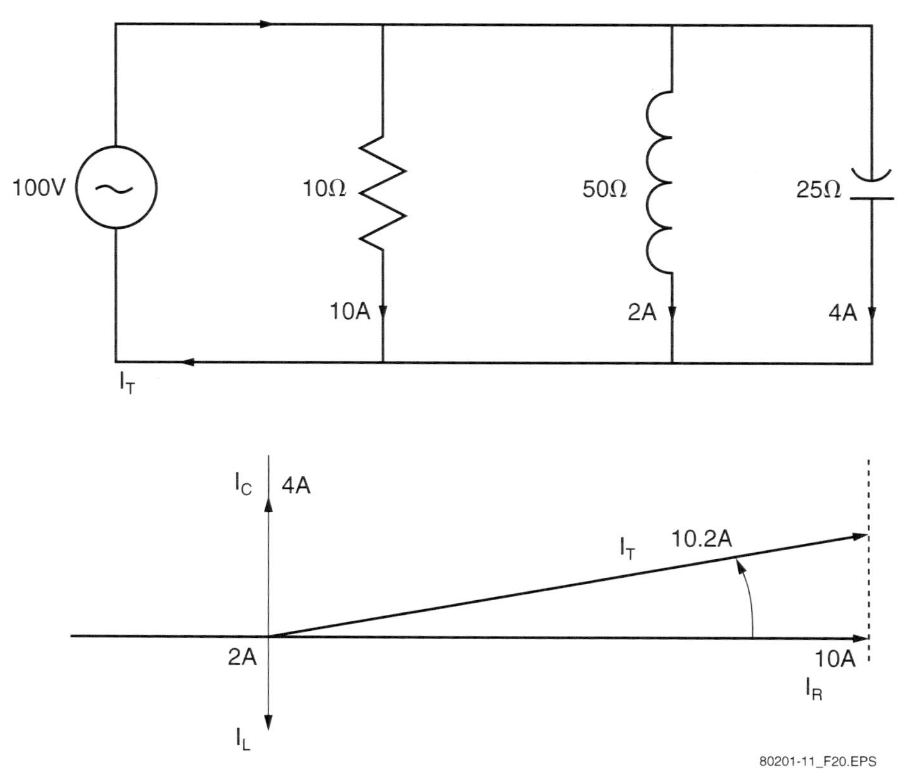

Figure 20 RLC circuit calculation.

On Site

Managing Power Factor

In hot weather, everyone turns on air conditioners. Air conditioners have compressors and fans that are driven by motors. These motors are inductive loads. The addition of so many inductive loads in the system can result in a poor system power factor and excessive power losses.

Banks of capacitors are automatically switched into the system so that their capacitive reactance cancels out as near as possible the inductive reactance presented by the motors. Remember that inductors shift the current and voltage away from each other in one direction (ELI), and capacitors shift them away from each other in the opposite direction (ICE). This maintains the system power factor at an acceptable value and keeps system losses to a minimum.

9.5.0 Power Triangle

The phase relationships among the three types of AC power are easily visualized on the power triangle shown in *Figure 21*. The true power (W) is the horizontal leg, the apparent power (VA) is the hypotenuse, and the cosine of the phase angle between them is the power factor. The vertical leg of the triangle is the reactive power and has units of volt-amperes-reactive (VAR).z

As illustrated on the power triangle (*Figure 21*), the apparent power is always greater than the true power or reactive power. Also, the apparent power is the result of the vector addition of true and reactive power. The power magnitude relationships shown in *Figure 21* can be derived from the Pythagorean theorem for right triangles:

$$c^2 = a^2 + b^2$$

Therefore, c also equals the square root of $a^2 + b^2$, as shown:

$$c = \sqrt{a^2 + b^2}$$

10.0.0 Transformers

A transformer is a device that transfers electrical energy from one circuit to another by electromagnetic induction (transformer action). The electrical energy is always transferred without a change in frequency, but the transfer may involve changes in the magnitudes of voltage and current. Because a transformer works on the principle of electromagnetic induction, it must be used with an input source voltage that varies in amplitude.

10.1.0 Transformer Construction

Figure 22 shows the basic components of a transformer. In its most basic form, a transformer consists of the following:

- Primary coil or winding
- Secondary coil or winding
- Core that supports the coils or windings

A simple transformer action is shown in *Figure 23*. The primary winding is connected to a 60Hz AC voltage source. The magnetic field or flux builds up (expands) and collapses (contracts) around the primary winding. The expanding and contracting magnetic field around the primary winding cuts the secondary winding and induces an alternating voltage into the winding. This voltage causes AC to flow through the load. The voltage may be stepped up or down depending on the design of the primary and secondary windings.

10.1.1 Core Properties

Commonly used core materials are air, soft iron, and steel. Each of these materials is suitable for particular applications and unsuitable for others. Generally, air-core transformers are used when the voltage source has a high frequency (above 20kHz). Iron-core transformers are generally used when the source frequency is low (below 20kHz). A soft-iron transformer is very useful where the transformer must be physically small yet efficient. The iron-core transformer provides better power transfer than the air-core transformer. Laminated sheets of steel are often used in a transformer to reduce one type of power loss known as eddy currents. These are undesirable currents, induced into the core, which circulate around the core. Laminating the core reduces these currents to smaller levels. These steel laminations are insulated with a nonconducting material, such as varnish, and then formed into a core as shown in *Figure 24*. It takes about 50 such laminations to make a core one-inch thick. The most efficient transformer core is one that offers the best path for the most lines of flux, with the least loss in magnetic and electrical energy.

10.1.2 Transformer Windings

A transformer consists of two coils called windings, which are wrapped around a core. The transformer operates when a source of AC voltage

Power Factor

GOING GREEN

In power distribution circuits, it is desirable to achieve a power factor approaching a value of 1 in order to obtain the most efficient transfer of power. In AC circuits where there are large inductive loads such as in motors and transformers, the power factor can be considerably less than 1. For example, in a highly inductive motor circuit, if the voltage is 120V, the current is 12A, and the current lags the voltage by 60 degrees, the power factor is 0.5 or 50 percent (cosine of 60 degrees = 0.5). The apparent power is 1,440VA (120V × 12A), but the true power is only 720W [120V × (0.5 × 12A) = 720W]. This is a very inefficient circuit. What would you do to this circuit in order to achieve a circuit having a power factor as close to 1 as possible?

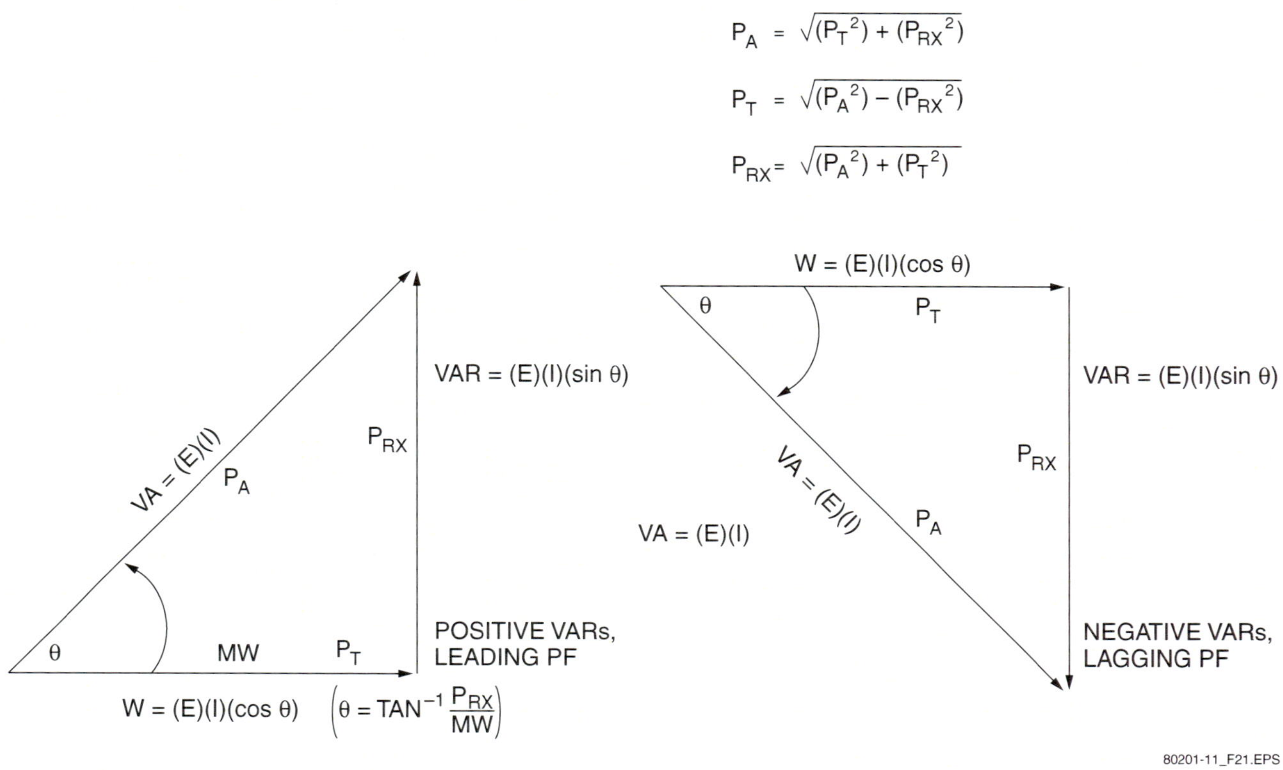

Figure 21 Power triangle.

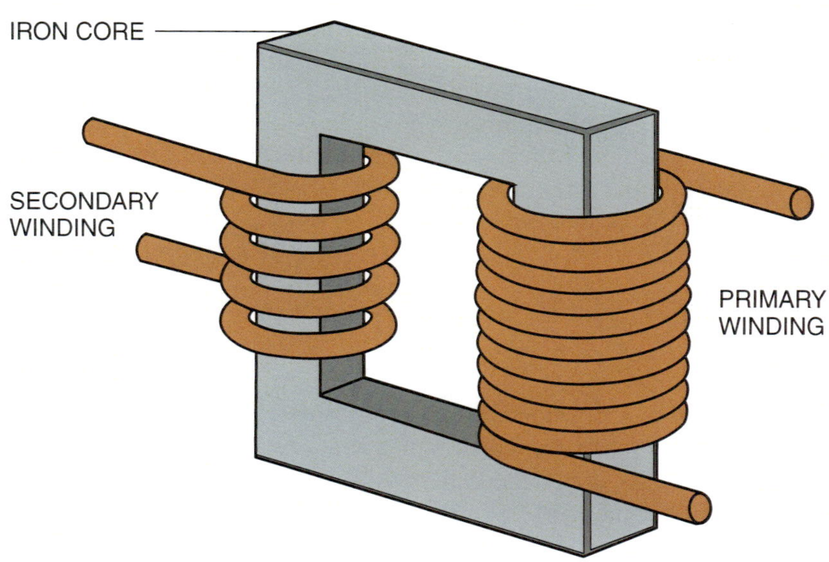

Figure 22 Basic components of a transformer.

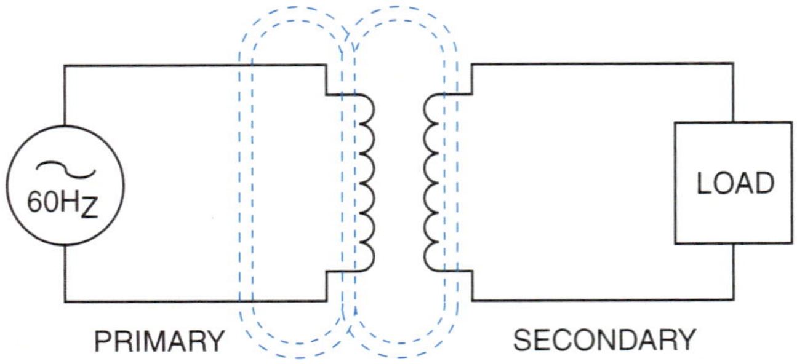

Figure 23 Transformer action.

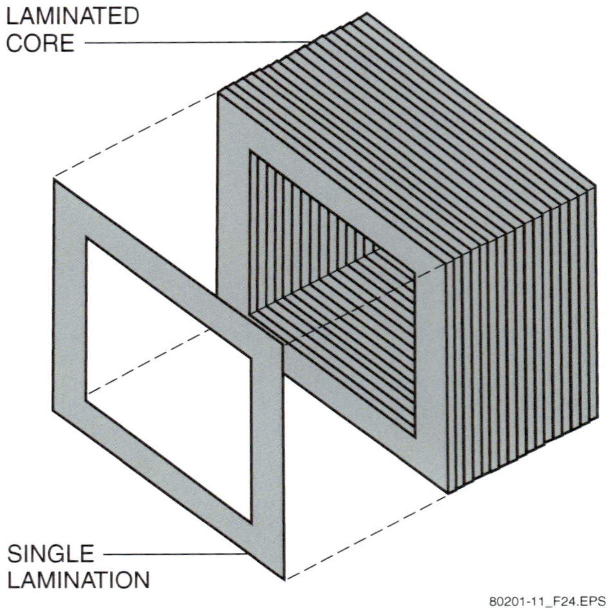

Figure 24 Steel laminated core.

is connected to one of the windings and a load device is connected to the other. The winding that is connected to the source is called the primary winding. The winding that is connected to the load is called the secondary winding. *Figure 25* shows a cutaway view of a typical transformer.

The wire is coated with varnish so that each turn of the winding is insulated from every other turn. In a transformer designed for high-voltage applications, sheets of insulating material such as

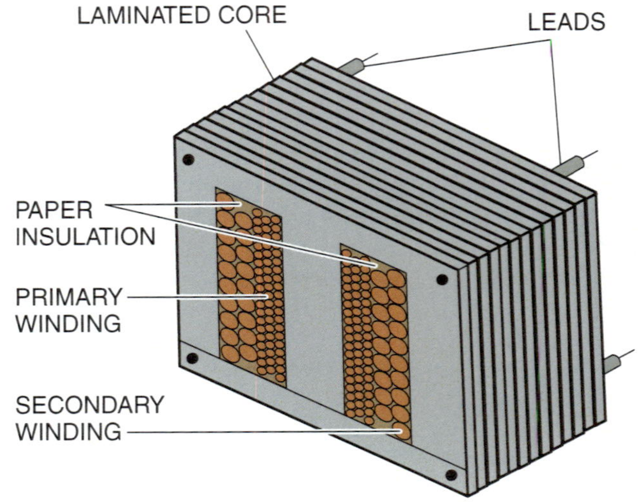

Figure 25 Cutaway view of a transformer core.

paper are placed between the layers of windings to provide additional insulation.

When the primary winding is completely wound, it is wrapped in insulating paper or cloth. The secondary winding is then wound on top of the primary winding. After the secondary winding is complete, it too is covered with insulating paper. Next, the core is inserted into and around the windings as shown.

Sometimes terminals may be provided on the enclosure for connections to the windings. *Figure 25* shows four leads, two from the primary and two from the secondary. These leads must be connected to the source and load, respectively.

10.2.0 Operating Properties

Regardless of type, most transformers operate in the same way. This section covers the no-load operation and phase relationships in a simple transformer.

10.2.1 Energized with No Load

A no-load condition is said to exist when a voltage is applied to the primary, but no load is con-

On Site

Transformers

Transformers are essential to all electrical systems and all types of electronic equipment. They are especially crucial to the operation of AC high-voltage power distribution systems. Transformers are used to both step up voltage and step down voltage throughout the distribution process. For example, a typical power generation plant might generate AC power at 13,800V, step it up to 230,000V for distribution over long transmission lines, step it down to 13,800V again at substations located at different points for local distribution, and finally step it down again to 240V and 120V for lighting and local power use.

nected to the secondary. Assume the output of the secondary is connected to a load by an open switch. Because of the open switch, there is no current flowing in the secondary winding. With the switch open and an AC voltage applied to the primary, there is, however, a very small amount of current, called exciting current, flowing in the primary. This current excites the coil of the primary to create a magnetic field. The amount of exciting current is determined by three factors: the amount of voltage applied (E_A); the resistance (R) of the primary coil's wire and core losses; and the X_L, which depends on the frequency of the exciting current. These factors are all controlled by transformer design.

This very small amount of exciting current serves two functions:

- Most of the exciting energy is used to support the magnetic field of the primary.
- A small amount of energy is used to overcome the resistance of the wire and core. This is dissipated in the form of heat (power loss).

Exciting current flows in the primary winding at all times to maintain this magnetic field, but no transfer of energy takes place as long as the secondary circuit is open.

10.2.2 Phase Relationship

The secondary voltage of a simple transformer may be either in phase or out of phase with the primary voltage. This depends on the direction in which the windings are wound and the arrangement of the connection to the external circuit (load). Simply, this means that the two voltages may rise and fall together, or one may rise while the other is falling. Transformers in which the secondary voltage is in phase with the primary are referred to as like-wound transformers, while those in which the voltages are 180 degrees out of phase are called unlike-wound transformers.

Dots are used to indicate points on a transformer schematic symbol that have the same polarity (points that are in phase). The use of phase-indicating dots is illustrated in *Figure 26*. In the first part of the figure, both the primary and secondary windings are wound from top to bottom in a clockwise direction, as viewed from above the windings. When built in this manner, the top lead of the primary and the top lead of the secondary have the same polarity. This is indicated by the dots on the transformer symbol.

The second part of the figure shows a transformer in which the primary and secondary are wound in opposite directions. As viewed from above the windings, the primary is wound in a clockwise direction from top to bottom, while the secondary is wound in a counterclockwise direction. Notice that the top leads of the primary and secondary have opposite polarities. This is indicated by the dots being placed on opposite ends of the transformer symbol. Thus, the polarity of voltage at the terminals of the transformer secondary depends on the direction in which the secondary is wound with respect to the primary.

10.3.0 Turns and Voltage Ratios

To understand how a transformer can be used to step up or step down voltage, the concept of turns ratio must be understood. The total voltage induced into the secondary winding of a transformer is determined mainly by the ratio of the number of turns in the primary to the number of turns in the secondary, and by the amount of voltage applied to the primary. Therefore, to set up a formula:

$$\text{Turns ratio} = \frac{\text{number of turns in the primary}}{\text{number of turns in the secondary}}$$

The first transformer in *Figure 27* shows a transformer whose primary consists of ten turns of wire, and whose secondary consists of a single turn of wire. As lines of flux generated by the primary expand and collapse, they cut both the ten turns of the primary and the single turn of the secondary. Since the length of the wire in the secondary is roughly the same as the length of the wire in each turn of the primary, the EMF induced into the secondary will be the same as the EMF induced into each turn of the primary.

This means that if the voltage applied to the primary winding is 10 volts, the EMF in the primary is almost 10 volts. Thus, each turn in the primary has an induced EMF of one-tenth of the total applied voltage, or one volt. Since the same flux lines cut the turns in both the secondary and the primary, each turn has an EMF of one volt induced into it. The first transformer in *Figure 27* has only one turn in the secondary, thus, the EMF across the secondary is one volt.

The second transformer in *Figure 27* has a ten-turn primary and a two-turn secondary. Since the flux induces one volt per turn, the total voltage across the secondary is two volts. Notice that the volts per turn are the same for both primary and secondary windings. Since the EMF in the primary is equal (or almost) to the applied voltage, a proportion may be set up to express the value of the voltage induced in terms of the voltage applied to the primary and the number of turns in each winding. This proportion also shows the relationship between the number of turns in each

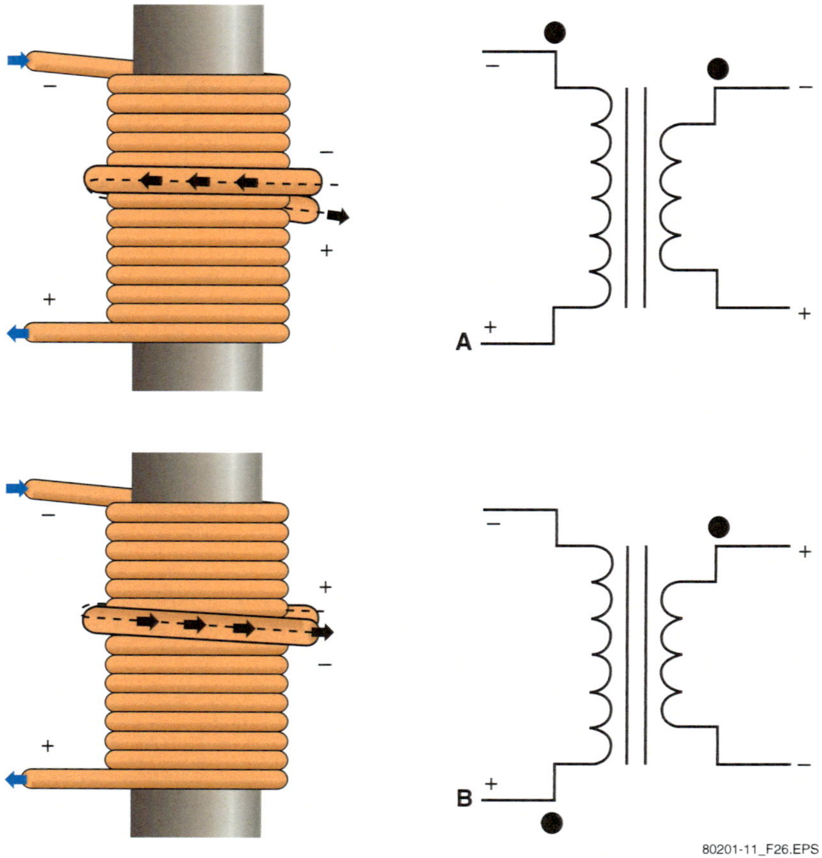

Figure 26 Transformer winding polarity.

winding and the voltage across each winding, and is expressed by the equation:

$$\frac{E_S}{E_P} = \frac{N_S}{N_P}$$

Where:

N_P = number of turns in the primary
E_P = voltage applied to the primary
E_S = voltage induced in the secondary
N_S = number of turns in the secondary

The equation shows that the ratio of secondary voltage to primary voltage is equal to the ratio of secondary turns to primary turns. The equation can be written as:

$$E_P N_S = E_S N_P$$

For example, a transformer has 100 turns in the primary, 50 turns in the secondary, and 120VAC applied to the primary (E_P). What is the voltage across the secondary (E_S)?

N_P = 100 turns
N_S = 50 turns
E_P = 120VAC

$$\frac{E_S}{E_P} = \frac{N_S}{N_P} \text{ or } E_S = \frac{E_P N_S}{N_P}$$

$$E_S = \frac{120V \times 50 \text{ turns}}{100 \text{ turns}} = 60VAC$$

The transformers in *Figure 27* have fewer turns in the secondary than in the primary. As a result, there is less voltage across the secondary than across the primary. A transformer in which the voltage across the secondary is less than the voltage across the primary is called a step-down transformer. The ratio of a 10-to-1 step-down transformer is written as 10:1.

A transformer that has fewer turns in the primary than in the secondary produces a greater voltage across the secondary than the voltage applied to the primary. A transformer in which the voltage across the secondary is greater than the voltage applied to the primary is called a step-up transformer. The ratio of a 1-to-4 step-up transformer should be written 1:4. Notice in the two ratios that the value of the primary winding is always stated first.

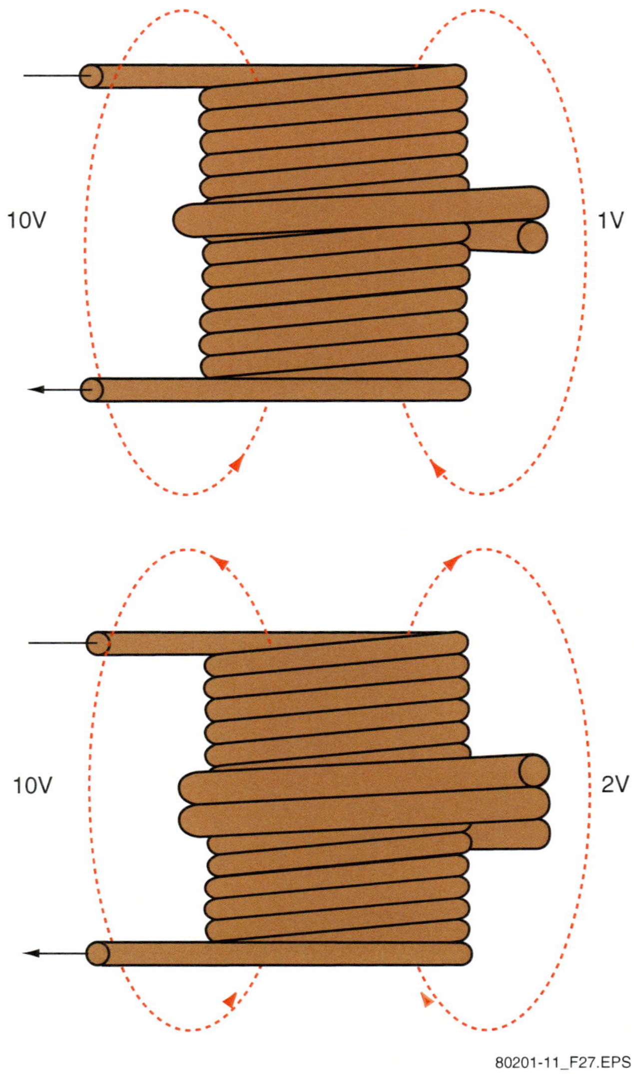

Figure 27 Transformer turns ratio.

10.4.0 Types of Transformers

Transformers are widely used to permit the use of trip coils and instruments of moderate current and voltage capacities and to measure the properties of high-voltage and high-current circuits. Since secondary voltage and current are directly related to primary voltage and current, measurements can be made under the low-voltage or low-current conditions of the secondary circuit and still determine primary properties. Tripping transformers and instrument transformers are examples of this use of transformers.

The primary or secondary coils of a transformer can be tapped to permit multiple input and output voltages. *Figure 28* shows examples of tapped transformers. The center-tapped transformer is particularly important because it can be used in conjunction with other components to convert an AC input to a DC output. There are many types of transformers, but the basic principles are the same for all transformers.

10.4.1 Isolation Transformer

Isolation transformers are wound so that their primary and secondary voltages are equal. Their purpose is to electrically isolate a piece of electrical equipment from the power distribution system.

Many pieces of electronic equipment use the metal chassis on which the components are mounted as part of the circuit (*Figure 29*). Personnel working with this equipment may accidentally come in contact with the chassis, completing the circuit to ground, and receive a shock, as shown

Think About It

Turns and Voltage Ratios

What is the magnitude of the voltage and current supplied by the secondary of the transformer in the circuit shown here?

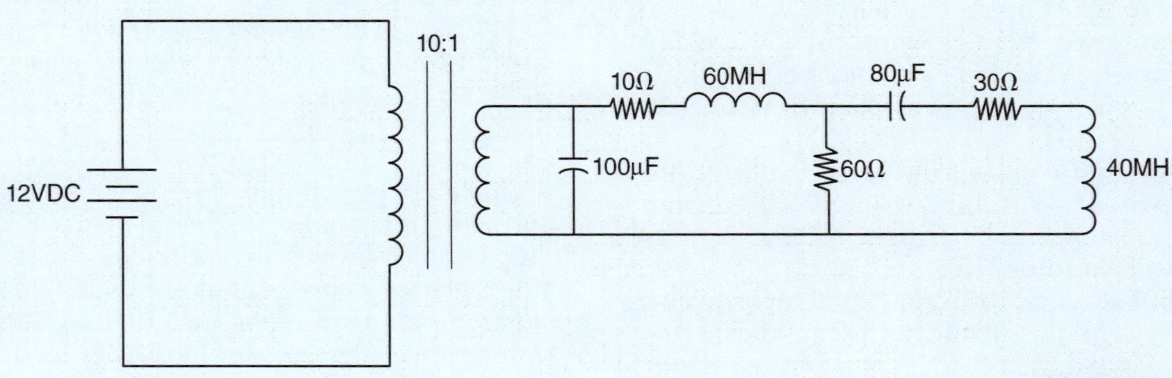

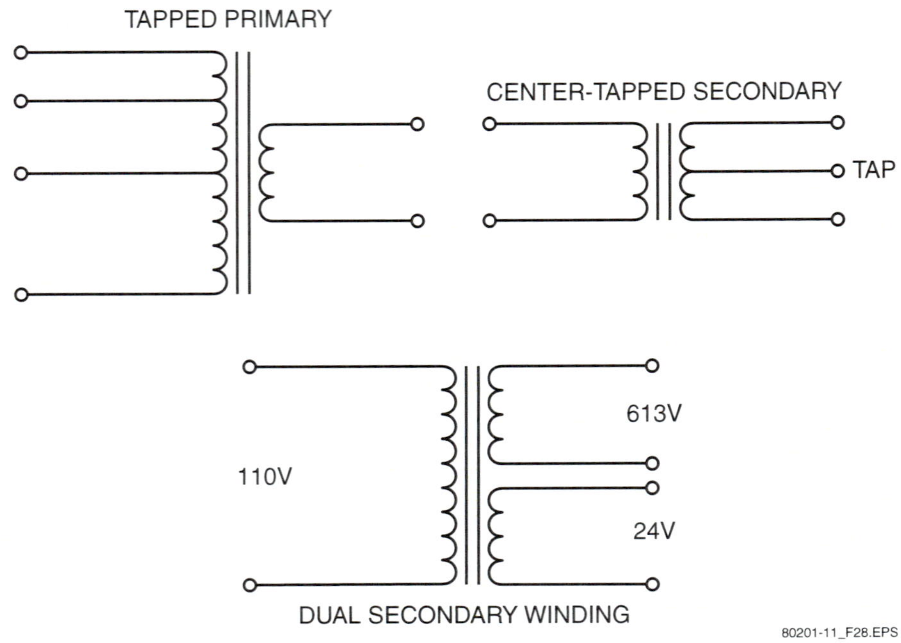

Figure 28 Tapped transformers.

in *Figure 29A*. If the resistances of a worker's body and the ground path are low, the shock can be fatal. Placing an isolation transformer in the circuit, as shown in *Figure 29B*, breaks the ground current path that includes the worker. Current can no longer flow from the power supply through the chassis and worker to ground; however, the equipment is still supplied with the normal operating voltage and current.

10.4.2 Autotransformer

In a transformer, the primary and secondary do not have to be separate and distinct windings. *Figure 30* is a schematic diagram of what is known as an autotransformer. Note that a single coil of wire is tapped to produce what is electrically both a primary and a secondary winding.

The voltage across the secondary winding has the same relationship to the voltage across the primary that it would have if they were two distinct windings. The movable tap in the secondary is used to select a value of output voltage either higher or lower than E_P, within the range of the transformer. When the tap is at Point A, E_S is less than E_P; when the tap is at Point B, E_S is greater than E_P.

Autotransformers rely on self-induction to induce their secondary voltage. The term *autotransformer* can be broken down into two words: *auto*, meaning self; and *transformer*, meaning to change potential. The autotransformer is made of one winding that acts as both a primary and a secondary winding. It may be used as either a step-up or step-down transformer. Some common uses of autotransformers are as variable AC voltage supplies and fluorescent light ballast transformers, and to reduce the line voltage for various types of low-voltage motor starters.

10.4.3 Current Transformer

A current transformer differs from other transformers in that the primary is a conductor to the load and the secondary is a coil wrapped around the wire to the load. Just as any ammeter is connected in line with a circuit, the current transformer is connected in series with the current to be measured. *Figure 31* is a diagram of a current transformer.

WARNING!

Do not open a current transformer under load. Although the use of a current transformer completely isolates the secondary and the related ammeter from the high-voltage lines, the secondary of a current transformer should never be left open circuited. To do so may result in dangerously high voltage being induced in the secondary.

Since current transformers are series transformers, the usual voltage and current relationships do not apply. Current transformers vary in rated primary current, but are usually designed with ampere-turn ratios such that the secondary delivers five amperes at full primary load.

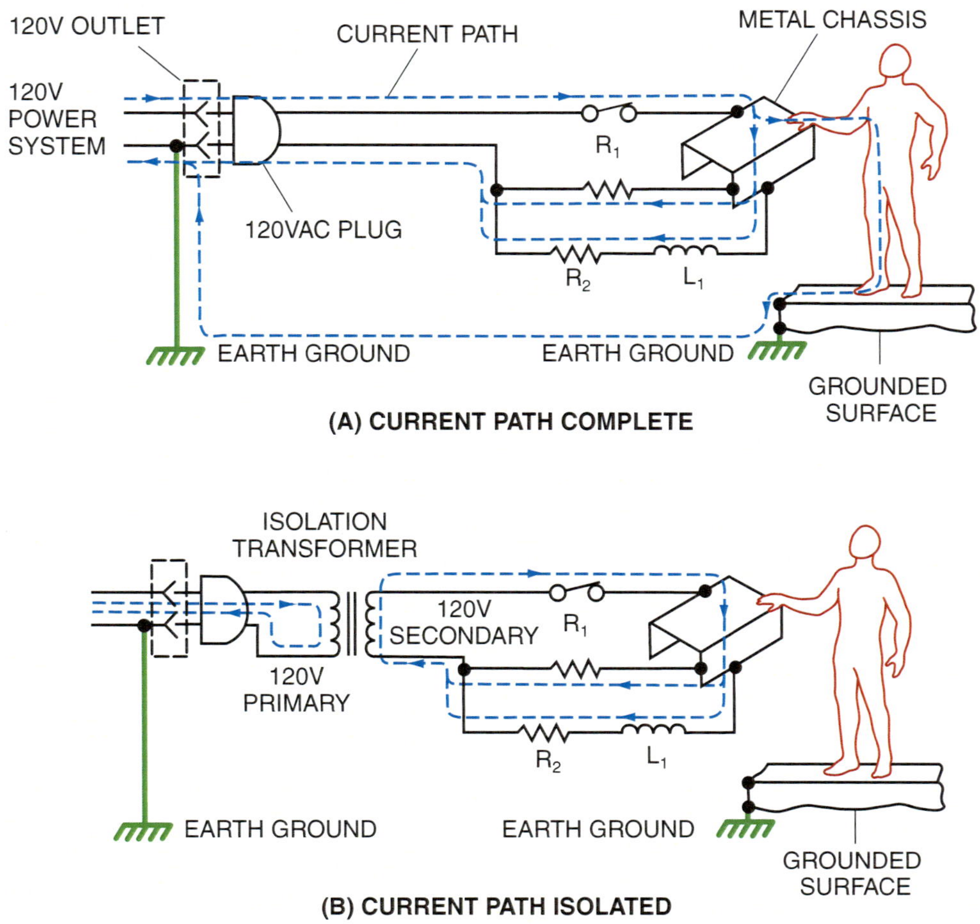

Figure 29 Importance of an isolation transformer.

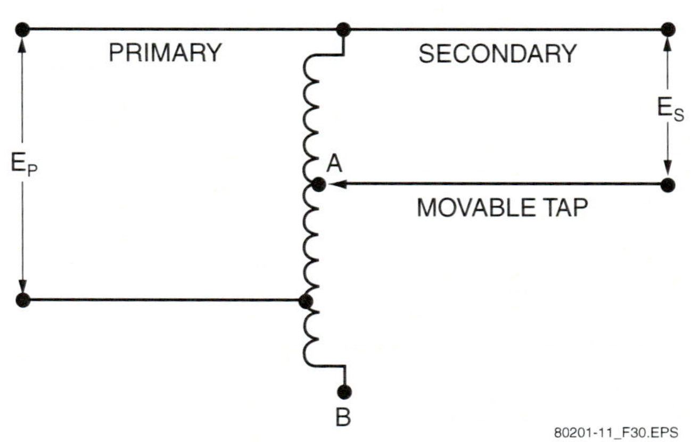

Figure 30 Autotransformer schematic diagram.

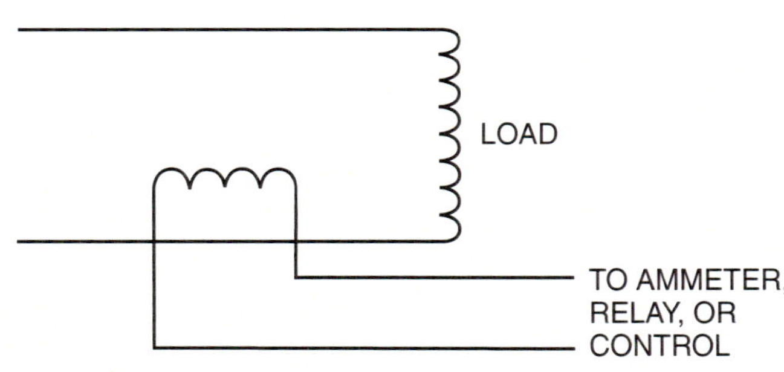

Figure 31 Current transformer schematic diagram.

Current transformers are generally built with only a few turns or no turns in the primary. The voltage in the secondary is induced by the changing magnetic field that exists around a single conductor. The secondary is wound on a circular core, and the large conductor that makes up the primary passes through the hole in its center. Because the primary has few or no turns, the secondary must have many turns (providing a high turns ratio) in order to produce a usable voltage. The advantage is that the output off the secondary is proportional to the current flowing through the primary, without much voltage drop across the primary. This is because the primary voltage equals the current times the impedance. The impedance is kept near zero by using no or very few primary turns. The disadvantage is that the secondary circuit cannot be opened with the primary energized. To do so would cause the secondary current to drop rapidly to zero. This would cause the magnetic field generated by the secondary current to collapse rapidly. The rapid collapse of the secondary field through the many

On Site

Isolation Transformers

In addition to being used to protect personnel from electrical shocks, shielded isolation transformers are widely used to prevent electrical disturbances on power lines from being transmitted into related load circuits. The shielded isolation transformer has a grounded shield between the primary and secondary windings that acts to direct unwanted signals to ground.

turns of the secondary winding would induce a high voltage in the secondary, creating an equipment and personnel hazard.

Because the output of current transformers is proportional to the current in the primary, they are most often used to power current-sensing meters and relays. This allows the instruments to respond to primary current without having to handle extreme magnitudes of current.

10.4.4 Potential Transformer

The primary of a potential transformer is connected across or in parallel with the voltage to be measured, just as a voltmeter is connected across a circuit. *Figure 32* shows the schematic diagram for a potential transformer.

Potential transformers are basically the same as any other single-phase transformer. Although primary voltage ratings vary widely according to the specific application, secondary voltage ratings are usually 120V, a convenient voltage for meters and relays.

The output of potential transformers is proportional to the phase-to-phase voltage of the primary. For that reason, they are often used to power voltage-sensing meters and relays. This allows the instruments to respond to primary voltage while having to handle only 120V. Also, potential transformers are essentially single-phase, step-down transformers. Therefore, power to operate low-voltage auxiliary equipment associated with high-voltage switchgear can be supplied off the high-voltage lines that the equipment serves via potential transformers.

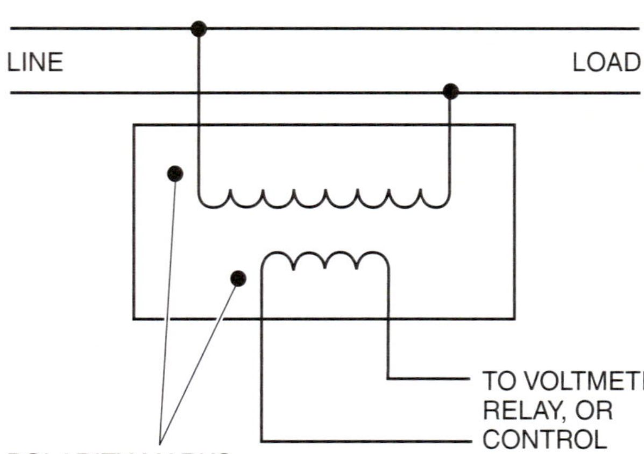

Figure 32 Potential transformer.

10.5.0 Transformer Selection

When replacing a defective transformer that is known to be original equipment, try to replace it with an exact duplicate. If it cannot be determined whether the transformer is original equipment or not, make sure that the replacement transformer safely and efficiently handles the primary voltage and current, and supports the voltage and current demands of the load.

Depending on its use, greater load demand properties may have to be considered when choosing a replacement control transformer. Three important properties are total steady-state VA (sometimes referred to as sealed VA), total inrush VA, and inrush load power factor. These three factors usually apply to transformers used with inductive circuits such as motor control circuitry, and transformers are used with electromagnetic control devices such as relays.

The total steady-state VA requirement is the rating that provides sufficient volt-amperes to the load in order to hold the load in a continuous energized state.

Electromagnetic control devices such as magnetic motor starter or relay coils may take anywhere from 20 to 60 milliseconds to become totally energized. The coil may draw up to 10 times the normal operating current during this time. This is referred to as the inrush VA and the transformer must be rated to handle these inrush current levels.

The true inrush load power factor cannot be determined without applying complex vector analysis to control transformer load components. Therefore, in determining the inrush load power factor, it is safe to apply a power factor of 40 percent. This means that from the transformer's available inrush power, only 40 percent is usable inrush power for the load. In other words,

On Site

Dangers of Working with AC Power

Working with AC power can be dangerous unless proper safety methods and procedures are followed. The National Institute for Occupational Safety and Health (NIOSH) investigated 224 incidents of electrocutions that resulted in occupational fatalities. One hundred twenty-one of the victims were employed in the construction industry, while 221 of the incidents (99 percent) involved AC. Of the 221 AC electrocutions, 74 (33 percent) involved AC voltages less than 600V and 147 (66 percent) involved 600V or more. Of the 74 lower-voltage electrocutions, 40 accidents involved 120/240V.

Factors relating to the causes of these electrocutions included the lack of enforcement of existing employer policies, including the use of personal protective equipment and the lack of supervisory intervention when existing policies were being violated. Of the 224 victims, 194 (80 percent) had some type of electrical safety training, while 39 victims had no training at all. It is notable that 100 of the victims had been on the job less than one year.

Never assume that you are safe when working at lower voltages. All voltage levels must be considered potentially lethal. Also, safety training is useless unless put it into practice every day. Always put safety first.

the transformer's total inrush VA rating must be about 160 percent of the inrush requirements of the loads if the total load consists of electromagnetic devices. Most replacement transformers have their inrush VA rating calculated for a power factor of 40 percent.

If an isolation transformer must be replaced, make sure to select a transformer designed for isolation. Its source and load impedance must match those of the original transformer. The replacement should also be designed to handle the signal levels and frequency range.

NOTE

Although an isolation transformer may have a 1:1 turns ratio, it should not be reversed in the circuit because a change in operating performance could result.

11.0.0 Three-Phase Power

Single-phase power is adequate to supply residences and small commercial businesses. In commercial and industrial uses, the power demand is greater, especially where large electric motors are used. Motors larger than one horsepower are usually three-phase motors.

The discussion of power generation has focused on a single conductor rotating in a magnetic field. If three rotating conductors are placed 120 degrees apart, three equal voltages are generated. As shown in *Figure 33*, they occur 120 degrees apart in time; in other words, they are 120 degrees out-of-phase with one another.

There are several ways in which three-phase power sources can be connected, depending on the voltage(s) and the amount of current required. Three-phase transformers selection is based on the voltage levels required by the customer(s). The transformers are rated for certain voltage ranges, typically 120V or 240V. The actual voltages produced depend on the way the transformers are connected. The four-wire wye-wye arrangement (*Figure 34*) is by far the most common in power distribution systems. It typically produces 120/208 volts or 277/480 volts. The 277 volts is used for fluorescent lighting. The delta-delta (closed delta) arrangement (*Figure 35*) can provide 120/240 volts or 240/480 volts. A delta-wye configuration (*Figure 36*) can produce 120/208V or 277/480 volts.

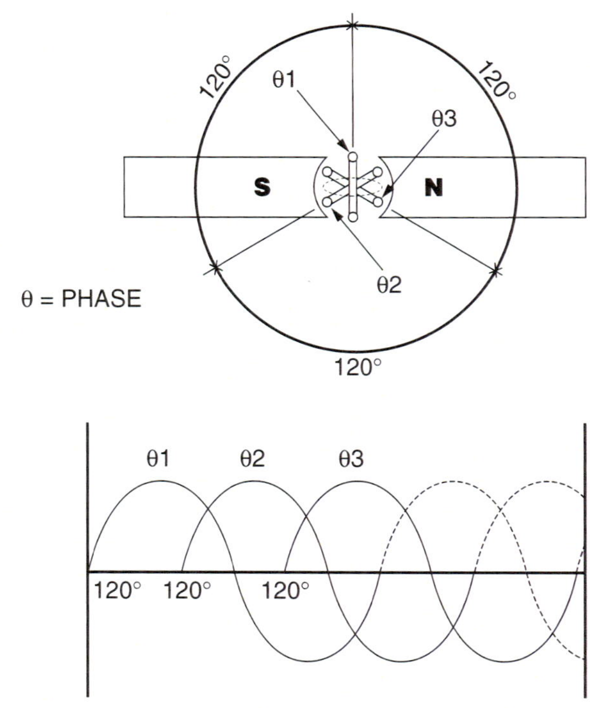

Figure 33 Three-phase voltage development.

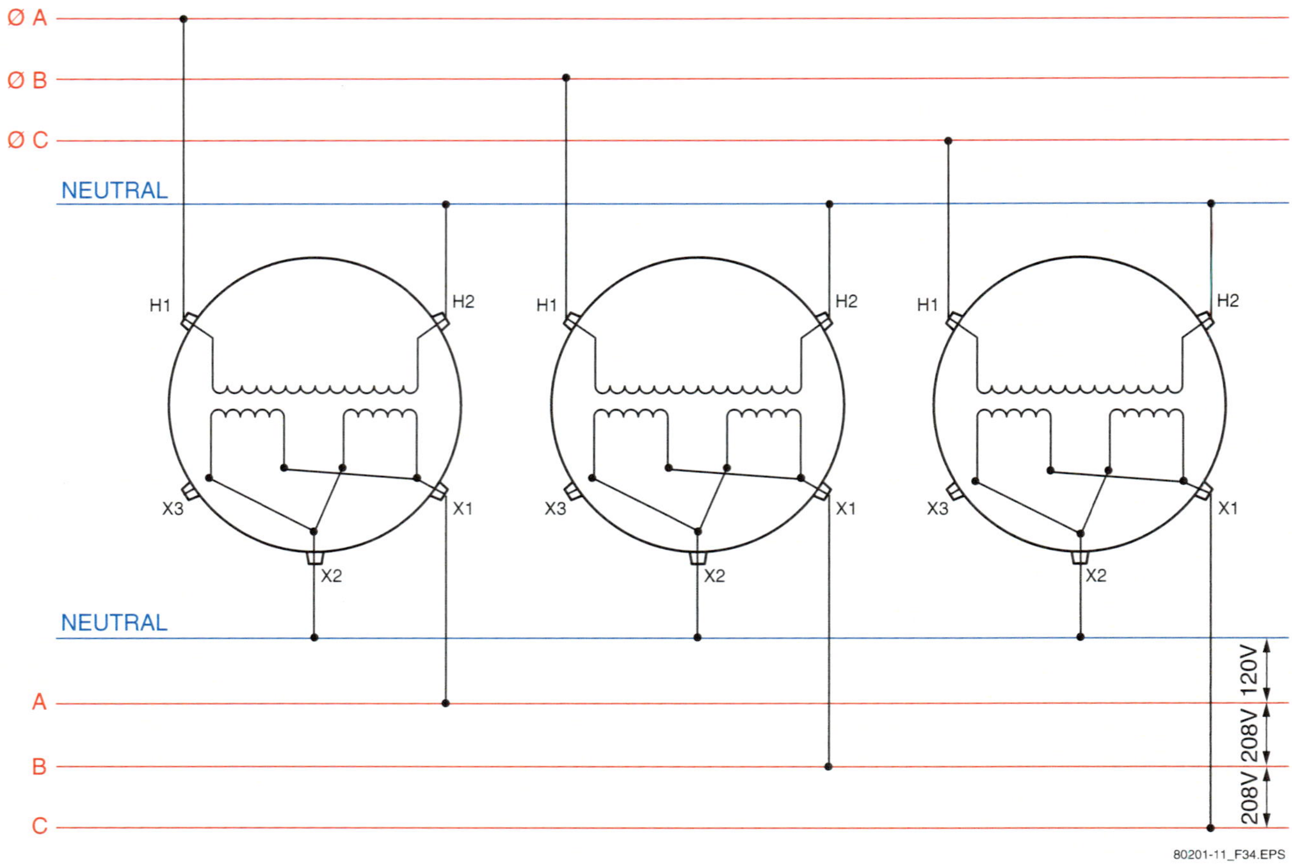

Figure 34 Wye-wye arrangement.

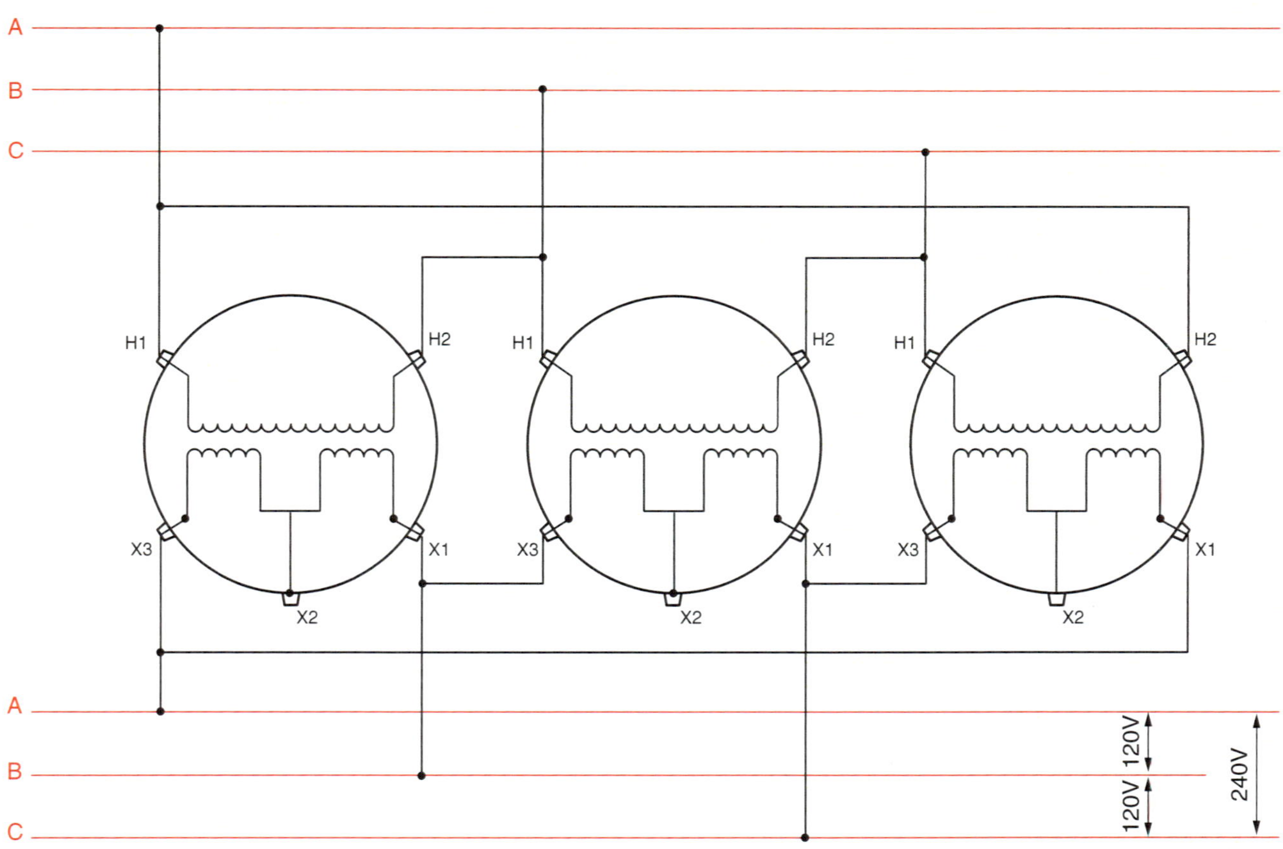

Figure 35 Closed delta arrangement.

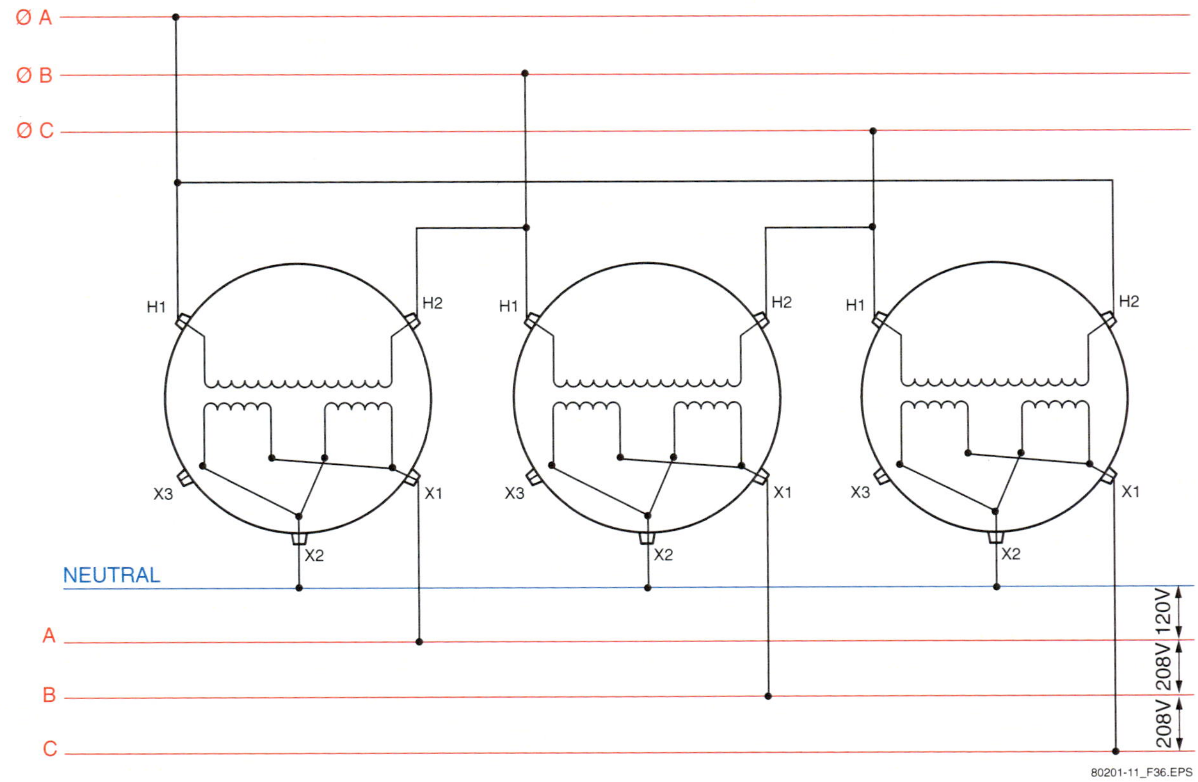

Figure 36 Delta-wye arrangement.

Primary voltages in power distribution systems typically carry three-phase power ranging from 2,400V to 34,000V. *Figure 37* shows an example of the transformer connections for a typical three-phase distribution system.

11.1.0 Voltage and Current Imbalance in Three-Phase Systems

In a three-phase system, a current imbalance in one leg can cause overheating in the other two legs. Three-phase systems therefore need to be checked periodically to make sure they are balanced. If the current is out of balance by more than 10 percent or the voltage is out of balance by more than 2 percent, the imbalance must be corrected. Sometimes the problem is at the source and must be corrected by the power company. This is a common occurrence following a power outage.

A small imbalance in the phase-to-phase voltages can result in a much greater current imbalance. With a current imbalance, the heat generated in motor windings and other inductive loads is increased. Both current and heat can cause nuisance overload trips and may cause motor/equipment failures. For this reason, the voltage imbalance between any two legs of the voltage applied to a three-phase motor or system should not exceed

On Site

Power Distribution

Many commercial and industrial operations need the higher voltages provided by three-phase power systems. Three-phase transformers provide that power, while single-phase transformers provide power to homes and small businesses. In modern residential subdivisions and business developments, power is routed underground and distributed from pad-mount transformers.

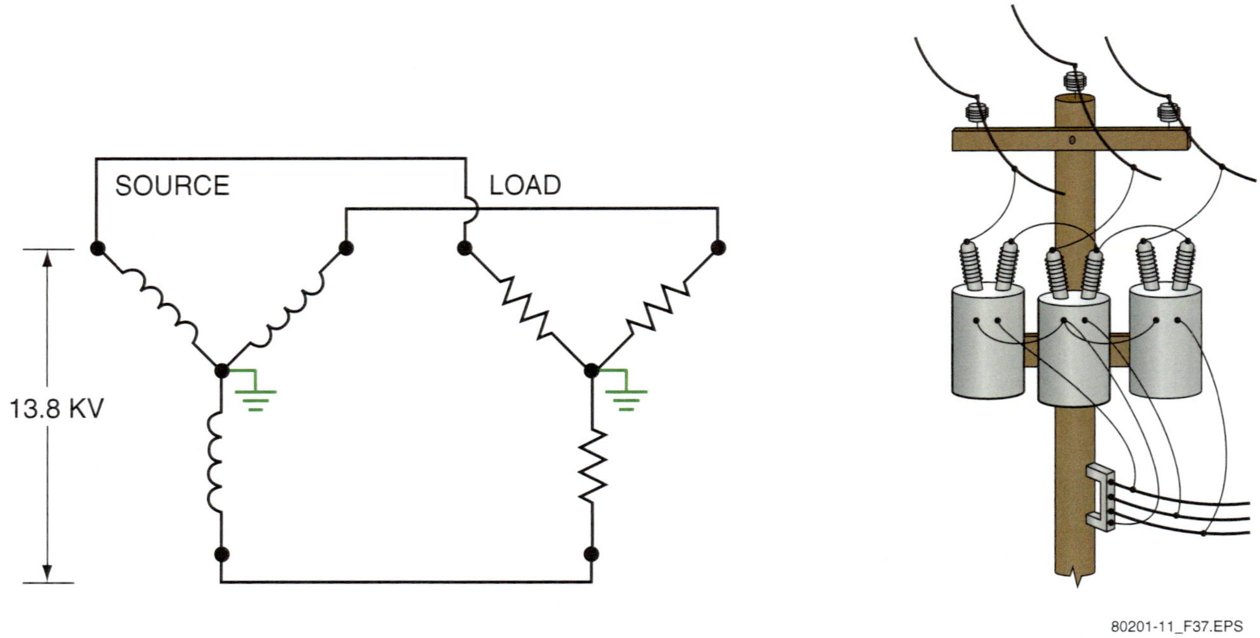

Figure 37 Three-phase transformer connections.

2 percent. If a voltage imbalance of more than 2 percent exists at the input to the equipment, the problem in the building or utility power distribution system should be corrected before operating the equipment.

Current imbalance between any two legs of a three-phase system should not exceed 10 percent. A current imbalance may occur without a voltage imbalance. This can happen when an electrical terminal, contact, etc., becomes loose or corroded, causing a high resistance in the leg. Since current follows the path of least resistance, the current in the other two legs will increase, causing more heat to be generated in the devices supplied by those legs.

Summary

The process by which current is produced electromagnetically is called induction. As the conductor moves across the magnetic field, it cuts the lines of force, and electrons within the conductor flow, creating an electromotive force (EMF).

Inductive and capacitance components offer impedance to current flow that is called reactance, rather than resistance. Inductive and capacitance reactances are out of phase with each other and with any DC resistive components in the circuit. Therefore, the total impedance to current flow must take the phase shift into account. In addition, reactive components do not consume power. Thus, the true power consumed by an AC circuit cannot be determined in the same way it is in a DC circuit.

Transformers are key components in AC circuits. They are used to provide circuit isolation and to increase (step up) or decrease (step down) AC voltage. The ratio of wire turns between the transformer primary and secondary determines the amount the voltage that will be stepped up or down.

Power is the amount of energy consumed by a circuit. In an AC circuit, a decrease in voltage occurs as the conductor intersects the magnetic field at an angle less than 90 degrees. The greatest current is produced when the conductor intersects the magnetic field at right angles to the flux lines.

Review Questions

1. An electric current always produces _____.
   a. mutual inductance
   b. a magnetic field
   c. capacitive reactance
   d. high voltage

2. The number of cycles that an alternating electric current undergoes per second is known as _____.
   a. amperage
   b. frequency
   c. voltage
   d. resistance

3. If the frequency of an AC current is 60 cycles per second, the hertz rating of the circuit is _____.
   a. 15 hertz
   b. 30 hertz
   c. 60 hertz
   d. 90 hertz

4. The period of a waveform is the _____.
   a. inverse of frequency
   b. square root of frequency
   c. sine of frequency
   d. square of frequency

5. The peak voltage in a 120VAC circuit is _____.
   a. 117 volts
   b. 120 volts
   c. 150 volts
   d. 170 volts

6. Which of the following conditions exists in a circuit containing pure resistance?
   a. The voltage and current are in phase.
   b. The voltage and current are 90 degrees out of phase.
   c. The voltage and current are 120 degrees out of phase.
   d. The voltage and current are 180 degrees out of phase.

7. In a purely resistive circuit where 240V is applied across a 10-ohm resistor, the amperage is _____.
   a. 10A
   b. 24A
   c. 60A
   d. 120A

8. Which of the following will cause the inductance of a coil to decrease?
   a. Adding turns of wire
   b. Decreasing coil diameter
   c. Decreasing turn spacing
   d. Using a magnetic core

9. The inductive reactance of a 15-henry inductor in a 60Hz circuit is _____.
   a. 2,132V
   b. 4,712V
   c. 5,652V
   d. 6,937V

10. Reactance increases with an increase in inductance.
    a. True
    b. False

11. Capacitance is measured in _____.
    a. farads
    b. joules
    c. henrys
    d. amps

12. The total capacitance of a series circuit containing a 60μF capacitor and a 40μF capacitor is approximately _____.
    a. 24μF
    b. 32μF
    c. 40μF
    d. 62μF

13. The total capacitance of a 30-microfarad capacitor in parallel with a 20-microfarad capacitor is _____.
    a. 20 microfarads
    b. 30 microfarads
    c. 50 microfarads
    d. 600 microfarads

14. The opposition to current flow offered by the capacitance of a circuit is known as _____.
    a. mutual inductance
    b. pure resistance
    c. inductive reactance
    d. capacitive reactance

15. Current leads the voltage by 90 degrees in a(n) _____ .
    a. capacitive circuit
    b. resistive circuit
    c. inductive circuit
    d. RC circuit

16. Which of the following conditions exist in a circuit of pure inductance?
    a. The voltage and current are in phase.
    b. The voltage and current are 90 degrees out of phase.
    c. The voltage and current are 120 degrees out of phase.
    d. The voltage and current are 180 degrees out of phase.

17. The total opposition to current flow in an AC circuit is known as _____.
    a. resistance
    b. capacitive reactance
    c. inductive reactance
    d. impedance

18. The formula $\sqrt{R^2 + X^2}$ is used to solve for the _____.
    a. total impedance of an AC circuit
    b. frequency of a parallel RC circuit
    c. total current in an AC circuit
    d. power factor of an AC circuit

19. True power is best described as the _____.
    a. power consumed by the combination of the resistive and reactive components of a circuit
    b. power consumed by the resistive components of a circuit
    c. power consumed by the inductors and capacitors in a circuit
    d. product of source voltage and current in an AC circuit

20. A power factor is not an angular measure, but a numerical ratio with a value between 0 and 1, equal to the _____.
    a. sine of the phase angle
    b. tangent of the phase angle
    c. cosine of the phase angle
    d. cotangent of the phase angle

21. The two windings of a conventional transformer are known as the _____.
    a. mutual and inductive windings
    b. high- and low-voltage windings
    c. primary and secondary windings
    d. step-up and step-down windings

22. A transformer has 300 turns on the primary, 100 turns on the secondary and 120VAC applied to the primary. The voltage across the secondary is _____.
    a. 33V
    b. 40V
    c. 90V
    d. 120V

23. A transformer with a primary of 200 turns and a secondary of 400 turns has a _____.
    a. step-up ratio of 4:2
    b. step-down ratio of 2:4
    c. step-up ratio of 1:2
    d. step-down ratio of 2:1

24. Inrush current is a factor in selecting transformers for _____.
    a. RC circuits
    b. resistive circuits
    c. capacitive circuits
    d. inductive circuits

25. In a three-phase system, problems are likely to occur if the voltage imbalance between any two legs is greater than _____.
    a. 2 percent
    b. 5 percent
    c. 8 percent
    d. 10 percent

Trade Terms Introduced in This Module

Capacitance: The storage of electricity in a capacitor; capacitance produces an opposition to voltage change. The unit of measurement for capacitance is the farad (F) or microfarad (μF).

Frequency: The number of cycles an alternating electric current, sound wave, or vibrating object undergoes per second.

Hertz (Hz): A unit of frequency; one hertz equals one cycle per second.

Impedance: The opposition to current flow in an AC circuit; impedance includes resistance (R), capacitive reactance (X_C), and inductive reactance (X_L). Impedance is measured in ohms.

Inductance: The creation of a voltage due to a time-varying current; also, the opposition to current change, causing current changes to lag behind voltage changes. The unit of measure for inductance is the henry (H).

Micro: Prefix designating one-millionth of a unit. For example, one microfarad is one-millionth of a farad.

Peak voltage: The peak value of a sinusoidally varying (cyclical) voltage or current is equal to the root-mean-square (rms) value multiplied by the square root of two (1.414). AC voltages are usually expressed as rms values; that is, 120 volts, 208 volts, 240 volts, 277 volts, 480 volts, etc., are all rms values. The peak voltage, however, differs. For example, the peak value of 120 volts (rms) is actually $120 \times 1.414 = 169.68$ volts.

Reactance: The imaginary part of impedance. Also, the opposition to alternating current (AC) due to capacitance (X_C) and/or inductance (X_L).

Resonance: A condition reached in an electrical circuit when the inductive reactance neutralizes the capacitance reactance, leaving ohmic resistance as the only opposition to the flow of current.

Root-mean-square (rms): The square root of the average of the square of the function taken throughout the period. The rms value of a sinusoidally varying voltage or current is the effective value of the voltage or current.

Self-inductance: A magnetic field induced in the conductor carrying the current.

Additional Resources

This module presents thorough resources for task training. The following resource material is suggested for further study.

Principles of Electric Circuits: Conventional Current Version, 2009. Thomas L. Floyd. New York: Prentice Hall.

Figure Credits

Topaz Publications, Inc., Module opener, Figure 13 (photo), and SA03
Fluke Corporation reproduced with permission, SA01

NCCER CURRICULA — USER UPDATE

NCCER makes every effort to keep its textbooks up-to-date and free of technical errors. We appreciate your help in this process. If you find an error, a typographical mistake, or an inaccuracy in NCCER's curricula, please fill out this form (or a photocopy), or complete the online form at **www.nccer.org/olf**. Be sure to include the exact module ID number, page number, a detailed description, and your recommended correction. Your input will be brought to the attention of the Authoring Team. Thank you for your assistance.

Instructors – If you have an idea for improving this textbook, or have found that additional materials were necessary to teach this module effectively, please let us know so that we may present your suggestions to the Authoring Team.

NCCER Product Development and Revision

13614 Progress Blvd., Alachua, FL 32615

Email: curriculum@nccer.org

Online: www.nccer.org/olf

❑ Trainee Guide ❑ AIG ❑ Exam ❑ PowerPoints Other ______________________________

Craft / Level: Copyright Date:

Module ID Number / Title:

Section Number(s):

Description:

____

____

____

____

Recommended Correction:

____

____

____

____

Your Name:

Address:

____

Email: Phone:

81201-11

Transmission Structure Rigging

Module Two

Trainees with successful module completions may be eligible for credentialing through NCCER's National Registry. To learn more, go to **www.nccer.org** or contact us at **1.888.622.3720**. Our website has information on the latest product releases and training, as well as online versions of our *Cornerstone* newsletter and Pearson's product catalog.

Your feedback is welcome. You may email your comments to **curriculum@nccer.org,** send general comments and inquiries to **info@nccer.org**, or use the User Update form at the back of this module.

V.1 11/11

81201-11
Transmission Structure Rigging

Objectives

When you have completed this module, you will be able to do the following:

1. Explain lift planning and rigging system planning.
2. Explain the strength calculations for various types of rigging including any derating for D/d ratios.
3. Explain crane stability as well as personnel platform requirements and operational considerations.
4. Demonstrate the determination of a load's weight and center of gravity.
5. Demonstrate the selection of an appropriate lifting device for a particular load.
6. Demonstrate hand signaling requirements for a blind lift.

Performance Tasks

Under the supervision of your instructor, you should be able to do the following:

1. Demonstrate the determination of a load's weight and center of gravity.
2. Demonstrate the selection of an appropriate lifting device for a particular load.
3. Demonstrate hand signaling requirements for a blind lift.

Trade Terms

Apex
Blind lift
D/d ratio

Industry Recognized Credentials

If you're training through an NCCER-accredited sponsor you may be eligible for credentials from NCCER's Registry. The ID number for this module is 81201-11. Note that this module may have been used in other NCCER curricula and may apply to other level completions. Contact NCCER's Registry at 888.622.3720 or go to www.nccer.org for more information.

Contents

Topics to be presented in this module include:

Figures and Tables

1.0.0 Introduction

This module covers complex rigging tasks that require advanced skills for the lifting and setting of transmission structures. The module also covers crane stability considerations during load lifts and detailed requirements for personnel platforms and their use.

Rigging projects come in all shapes, weights, and movement directions. Rigging activities for the assembly and setting of transmission structures may involve cranes or helicopters as well as a combination of rigging gear. For instance, it may take more than one crane to safely lift and set an assembled transmission structure. The use of helicopters or two cranes is a specialized application of rigging that requires qualified helicopter or crane operators and rigging personnel to work together.

Every lifting device has capacity and operating limits that must be observed for safe use. Always follow the manufacturer's instructions for use to prevent severe or fatal injuries.

2.0.0 Lift Planning and Execution

Before any work is performed, the personnel responsible for rigging should determine the following factors for lifts that deal with the assembly or setting of a particular type of transmission structure:

- Is the work site accessible by cranes or is a helicopter required?
- What is to be accomplished by each lift?
- What types of rigging equipment are required for each lift?
- Does the lifting and rigging equipment have the capacity to handle the required loads and forces?
- How can the hookups be made?
- What will happen when each load is first moved?
- What will be the travel path for each load to reach a desired destination?
- How will the each load be positioned or set down at the destination?
- Will the lift be a **blind lift**?
- How many personnel are needed to control each load safely during the moving, positioning, or setting of each load?
- What other lift factors must be considered, such as weather conditions, other electrical wires, clearance to the load destination, and visibility?

In some cases, a formal lift plan may be required to comply with any governmental authorities that have jurisdiction, especially for any blind lifts. Once these questions are answered and a lift plan is developed (if required), the lift can be performed, employing audible signals or hand signals provided by rigging personnel. Refer to the *Appendix* for illustrations of hand signals used in lifting operations.

3.0.0 Rigging System Planning

Lifting a load using rigging equipment requires determining the weight of the load, its center of gravity, and the vertical/horizontal force components that may affect the load.

3.1.0 Determining Load Weight

Determining below-the-hook load weight is essential for the proper selection of rigging and lifting equipment. Below-the-hook weight consists of the weight of the loads to be rigged as well as the weight of the rigging equipment used for lifting. Below-the-hook weight can be determined from one or more of the following sources:

- Shipping papers
- Manufacturer's information marked on the load, attached to the load, or available from the manufacturer
- Catalogs or construction drawings
- Tables of weights from manufacturers or handbooks

WARNING!

All of the below-the-hook weights must be provided to the crane operator so that the correct information from the crane load charts is used during the lift.

3.2.0 Determining Load Center of Gravity

WARNING!

A crane hook must always be above a load's center of gravity.

A load's center of gravity (CG) is the point at which the load balances. Its weight is considered concentrated at this balance point. When lifted, a load moves until the CG is directly below the point of support. To prevent this, the CG must always be located below the hoisting hook; otherwise the load will not be stable (*Figure 1*). When raised, an unstable load may tip or topple over, creating a hazard for personnel and equipment. Sling attachment points to the load must also be above the CG except

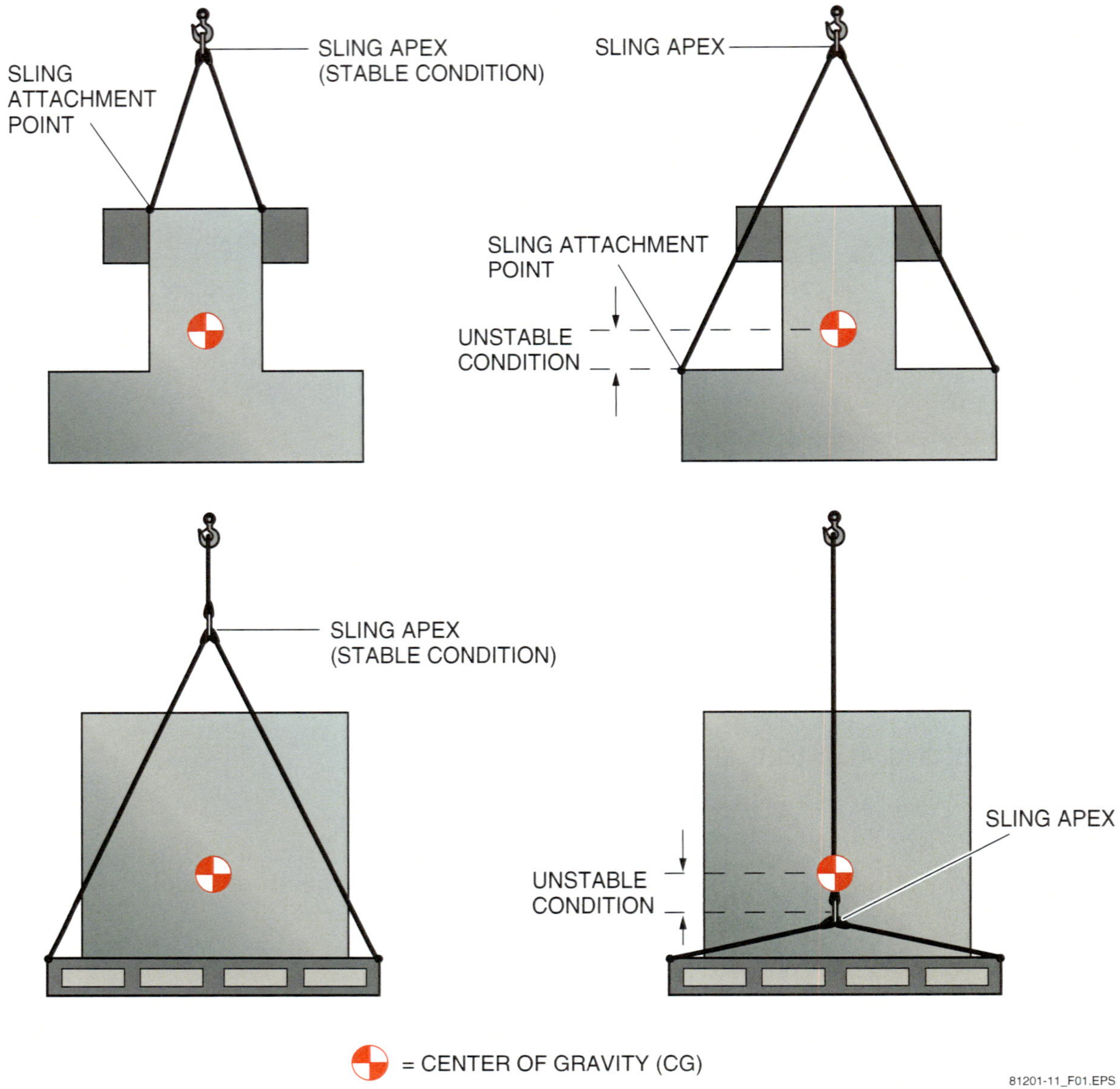

Figure 1 Load rigging stability.

when lifting platform-supported loads. When lifting platform-supported loads, the sling's **apex** must be above the CG, and the CG must be centered on the platform. The CG of a load may be determined by checking for marking or lifting lugs provided by the manufacturer on the load or documented in catalogs and manufacturer drawings. If the CG is not documented or marked, perform the following:

- Visually estimate the location of the load's CG.
- Select the type of rigging configuration that best accommodates the shape of the load and allows positioning of the crane hook over the CG.
- Make sure that the rigging configuration is strength-rated for the load weight as described in this module.
- Take up the slack in the rigging and the crane's load line.
- Lift the load just enough to check stability.
- If stable, continue the lift; if unstable, lower the load and adjust the rigging so that the lift point of the crane hook is closer to the point of load dip.
- Repeat the above until the load is stable.

WARNING!

If the shape of the load is not symmetrical, adjustment of the sling and/or beam lift rigging may require shortening or lengthening of sling legs and/or beam connection points. This helps shift the lift point to accommodate the load's CG. In this case, the strength of the sling legs must be recalculated as described for asymmetrical loads in this module.

4.0.0 Rigging Equipment Selection

Besides an appropriately load-rated crane or helicopter, the rigging equipment used for transmission structure erection or assembly is normally comprised of standard wire rope leg slings, or standard/custom spreader/lifting beams. Determining the proper strength of slings is one of the important tasks of personnel involved in rigging.

4.1.0 Slings

Some type of sling is used in nearly every lift. *Figure 2* shows the positioning of a lattice transmission tower bridge section using a four-legged sling. It is important to know how to select the correct sling for each situation, how to properly use the sling, and how to apply derating factors to compensate for sling angles.

Figure 3 illustrates the ways in which slings are measured. Note especially that the overall length of a sling does not take into account the diameters of the wire rope or any special fittings at either end. Note also that the width of the eye of a sling is measured from inside to inside at the widest point and that the width of all fittings is also measured from inside to inside.

The following are recommendations for selecting and using slings:

- Keep written records.
- Have a competent person inspect slings prior to and during their use.
- Check that proper rating tags are attached to any chains and master links.
- Use, test, maintain, and inspect all slings according to *29 CFR 1910.184*.
- Protect slings from sharp surfaces.
- Use a shackle when more than one sling is used at the same attachment point.
- Do not drag slings on abrasive surfaces.
- Start and stop slowly—shock loading is prohibited.
- Lift the load a few inches, and then check the rigging.

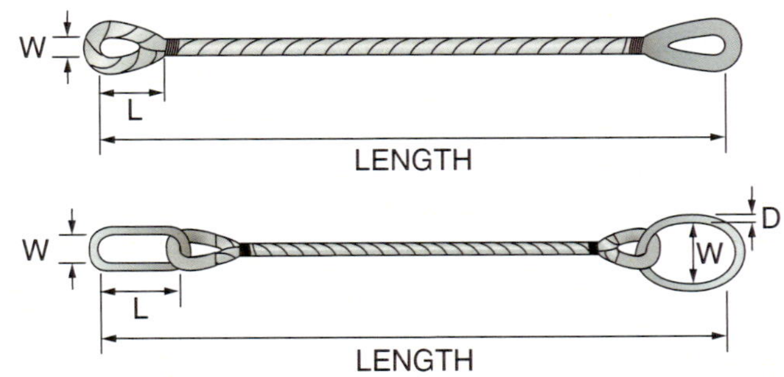

Figure 3 Sling measurement.

Figure 2 Setting the bridge on a transmission tower.

- Do not pull slings from under the load when the load is resting on the sling.
- Select a hitch that will hold and control the load.
- Allow for low **D/d ratios** as described in this module.
- Secure all unused sling legs when using more than one on a hook.
- Allow for increased tension caused by sling angles.
- Allow for reductions when using choker hitches.
- Allow for sling angles when forcing (seating) a choker or turning a load.
- Do not use wire rope clips to form eyes in slings.
- Keep fingers and hands clear of loads about to be lifted.
- Wear gloves when handling wire rope slings.

4.1.1 Sling Tensions

The sling angle is the angle formed by the leg of a sling with respect to horizontal when tension is put on the load. When using sling angle charts, it is important to know which angle is measured. Because the sling angle directly affects the rated capacity of any sling, proper sling angles are crucial to safe rigging (*Figure 4*).

A sling with two legs used to lift a 2,000-pound object will carry almost a 1,000-pound load in each leg if the sling angle is very close to 90 degrees to the horizontal. As the sling angle changes, the amount of load carried by each leg increases until, at 30 degrees to the horizontal, each leg of the sling is carrying the full weight of the load. Angles less than 30 degrees to the horizontal causes the legs of the sling to bear more than the load actually weighs. This will result either in excessive wear of the sling or in total failure of the sling.

Sling angles should be kept to greater than 45 degrees to horizontal, if possible. Optimum sling angles fall between 90 and 45 degrees to the horizontal. Angles of 30 degrees or less to the horizontal should be considered hazardous.

Table 1 is an example of a sling capacity chart. This type of chart is an effective tool for determining the safe working load (SWL) of a sling. The exact capacity of a sling can be determined from such tables by determining the sling wire size, classification, hitch configuration, and sling angle.

Nobody is expected to memorize long tables of the SWL, sometimes called the working load limit (WLL), for various configurations of slings. Such information is readily available from a variety of sources. There are, however, rules of thumb to determine the safety of a sling configuration.

> **NOTE**
> Pocket references such as *Bob's Rigging and Crane Handbook* are very useful for quickly calculating sling loads, as well as a wide variety of other rigging data.

Always refer to the manufacturer specifications when determining SWL. Should the need arise in the field to make a quick determination of SWL for a vertical choker hitch, the following formula can be used: first, convert the diameter of the wire

ANGLE FACTORS

ANGLE*	LOAD FACTOR
90 DEGREES	1.000
80 DEGREES	1.015
70 DEGREES	1.064
60 DEGREES	1.155
50 DEGREES	1.305
40 DEGREES	1.555
30 DEGREES	2.000

* SLING ANGLES ARE MEASURED FROM HORIZONTAL.

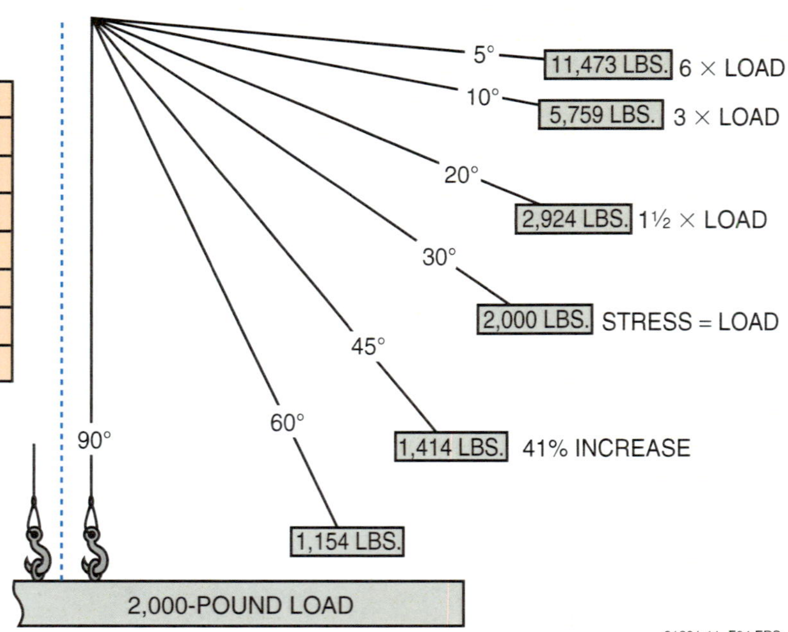

Figure 4 Sling angle factors.

Table 1 Example of a Sling Capacity Chart

Mechanical Splice in Pounds

Wire Rope EIPS/IWRC

Size in Inches			90°	60°	45°	30°
¼	1,300	960	2,600	2,200	1,820	1,300
5⁄16	2,000	1,480	4,000	3,400	2,800	2,000
⅜	2,800	2,200	5,600	5,000	4,000	2,800
7⁄16	3,800	2,800	7,600	6,800	5,400	3,800
½	5,000	3,800	10,000	8,800	7,200	5,000
9⁄16	6,400	4,800	12,800	11,000	9,000	6,400
⅝	7,800	5,800	15,600	13,600	11,000	7,800
¾	11,200	8,200	22,400	19,400	15,800	11,200
⅞	15,200	11,200	30,400	26,000	22,000	15,200
1	19,600	14,400	39,200	34,000	28,000	19,600
1⅛	24,000	18,000	48,000	42,000	34,000	24,000
1¼	30,000	22,500	60,000	52,000	42,000	30,000
			MULTIPLIER ⇨	1.00	0.75	0.60

Sling Length = Total distance between pick points × multiplier.

81201-11_T01.EPS

rope to eighths; then square the eighths and multiply by 150. The answer is a conservative SWL of the wire rope when choked. Carbon shackle sizes should exceed the wire rope size by ⅛" or more.

The SWL rule-of-thumb formulas for slings given in the following sections are based on the SWL of a single vertical hitch sling. It is highly recommended that riggers familiarize themselves with this type of configuration. The SWL of the single vertical hitch is based on the diameter of the wire rope and the type of wire rope being used. If the breaking strength and design (safety) factor of a wire rope is known from the manufacturer's data, the vertical hitch SWL can be obtained by dividing the breaking strength by the design factor. For wire rope that meets *ASME B30.9* requirements, the design factor is normally five. Every rigger should have an idea of the capacities of different sizes and types of wire rope. For any sling, the calculated SWL should always be greater than the load weight.

For wire rope slings that are wrapped around the load or hook, the D/d ratio for wire rope is important. It refers to the relationship between the diameter of the surface (D) the sling is around and the diameter of the sling wire rope (d). This relationship, specifically the severity of the bends in the sling, has an impact on the capacity of the sling. Efficiency charts have been developed to help determine the actual sling capacity. *Figure 5* shows load and hook diameter and a D/d ratio chart.

For example, consider a 1-inch-diameter rope sling in a basket hitch around a 2-inch shackle pin. The capacity of the sling in the basket hitch is listed as 17 tons. Look across the bottom of the chart for the 2-inch pin size and find the line that meets the ratio curve. It is opposite 65 percent efficiency. Multiply the rated load of 17 tons by 65 percent and you will get an actual capacity of 11 tons. Sling eyes with pin or body sizes that are at least equal to the rope diameter should be used with shackles and hooks.

4.1.2 Bridle Hitches

To determine the SWL for a selected size wire rope bridle hitch that is used to lift a symmetrical load, measure the length of the sling legs (L) and the distance between the hook and the center of the load (H). W is the rated/calculated SWL for one of the wire rope legs in a vertical pull. *Figure 6* shows a bridle hitch arrangement. When this information is obtained, use the following formula:

$$\text{SWL} = \text{W}(\text{H} \div \text{L})\text{No. of legs}$$

This formula is for a bridle hitch lifting a symmetrical load where the length and wire rope type/size of each leg is identical.

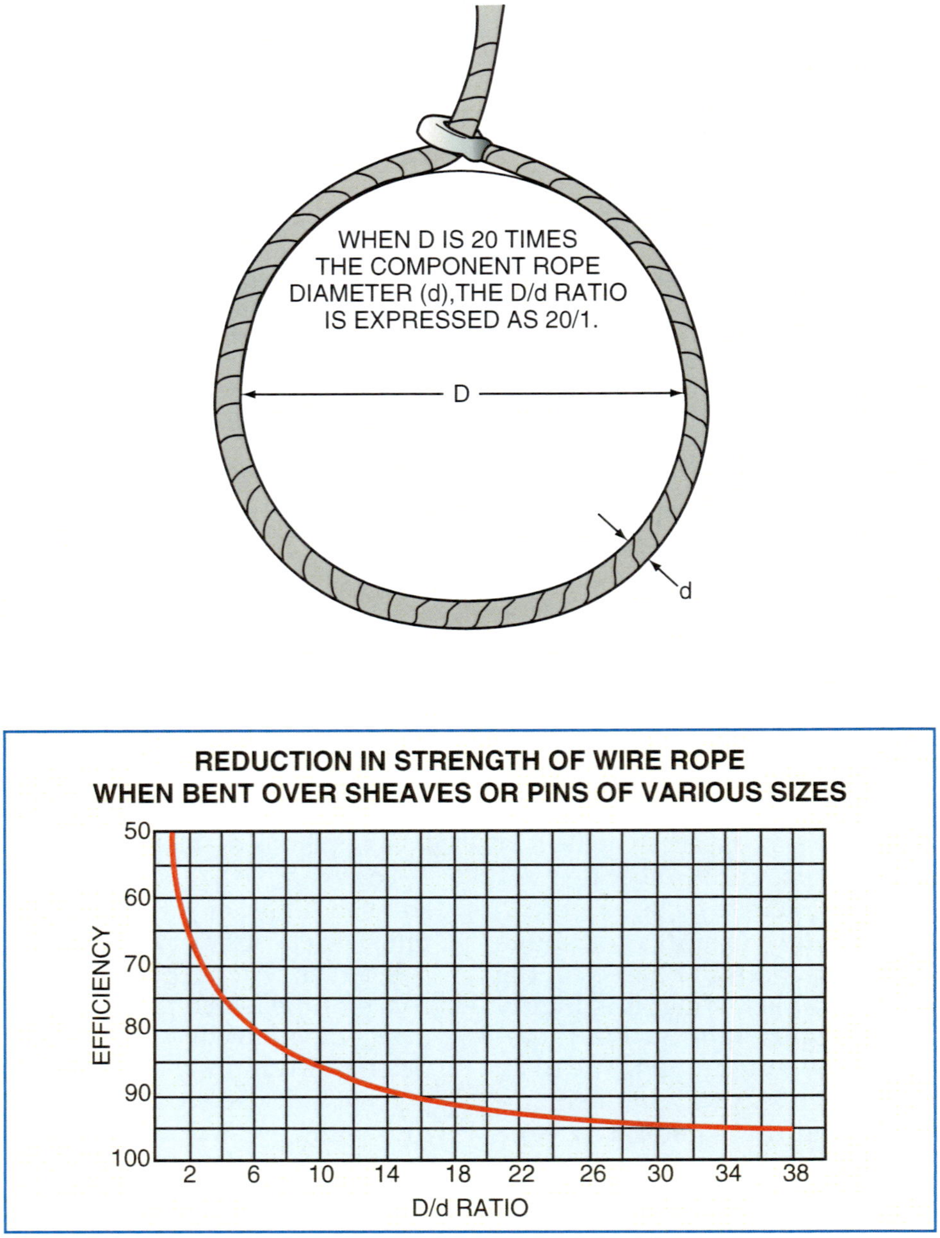

Figure 5 Load and hook diameter and D/d ratio chart for wire rope.

Rigging an asymmetrical load requires a much more precise formula to calculate sling stresses. As shown in *Figure 7,* a 24,000-pound unequal leg angular load is rigged with two slings with the hook positioned near an estimated load center of gravity. The length of Sling A is 29.7 feet, and the length of Sling B is 12 feet. The distance from the Sling A attachment point to directly below the lifting hook (D1) is 19 feet. The distance from the Sling B attachment point to directly below the lifting hook (D2) is 5 feet, and the height from the Sling A attachment point to the lifting hook (H1) is 22.8 feet. The height from Sling B attachment point to the lifting hook (H2) is 10.9 feet. Entering the numbers into the formula as shown in *Figure 7,* the load stress on Sling A is 11,099 pounds, and the load stress on Sling B is 17,041 pounds. The slings selected for asymmetrical loads should be based on a 6:1 safety factor instead of the normal 5:1 factor, to account for unexpected stresses.

4.1.3 Basket Hitches

Basket hitches include single, double, and double-wrapped arrangements. The following guidelines are used to determine the SWL of basket hitches:

- *Single basket hitch* (*Figure 8*) – For vertical legs, take the SWL of the single vertical hitch and multiply it by 2. For inclined legs, the formula is the same as that for the bridle hitch.

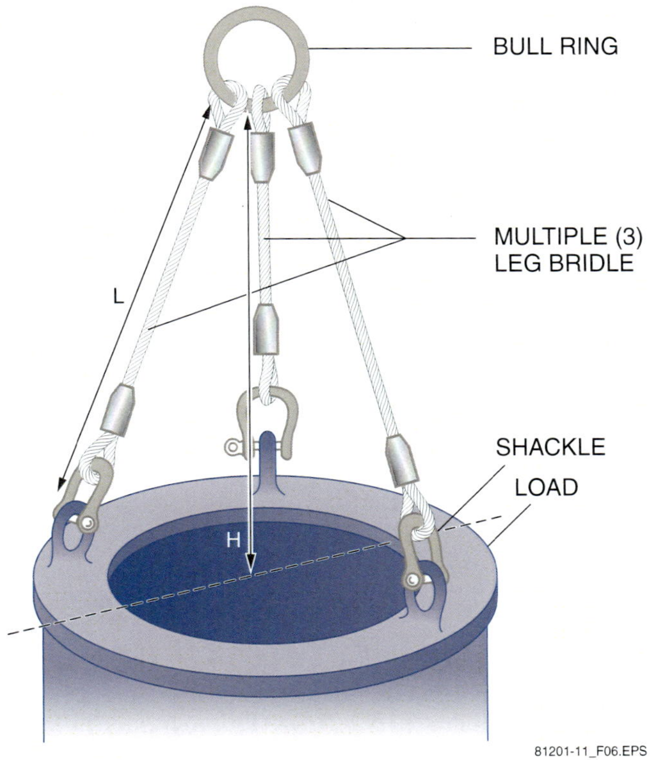

Figure 6 Bridle hitch.

- *Double basket hitch* – For vertical legs, multiply the SWL of the single vertical hitch by 4. For inclined legs, the formula is the same as that for the bridle hitch.

> **WARNING!**
>
> The rigger must verify that all four legs carry the load equally. Consider all D/d ratios.

- *Double-wrapped basket hitch* – The SWLs for double-wrapped basket hitches are the same as those for the double basket hitches.

4.1.4 Choker Hitches

Choker hitches include single leg, double leg, and double-wrapped arrangements. The two legs forming the choke naturally tighten to an approximate 45-degree sling angle inside the choke (*Figure 9*). Always refer to manufacturer data for the SWL of choker hitches. The following guidelines are used to determine the SWL of choker hitches:

ASYMMETRICAL BRIDLE SLING FORMULA

A MUCH MORE ACCURATE FORMULA TO CALCULATE SLING STRESS IS:

$$\text{SLING A} = \frac{\text{LOAD} \times D_2L_1}{D_2H_1 + D_1H_2}$$

$$\text{SLING B} = \frac{\text{LOAD} \times D_1L_2}{D_2H_1 + D_1H_2}$$

$$T_1 = \frac{\text{LOAD} \times D_2L_1}{D_2H_1 + D_1H_2}$$

$$T_2 = \frac{\text{LOAD} \times D_1L_2}{D_2H_1 + D_1H_2}$$

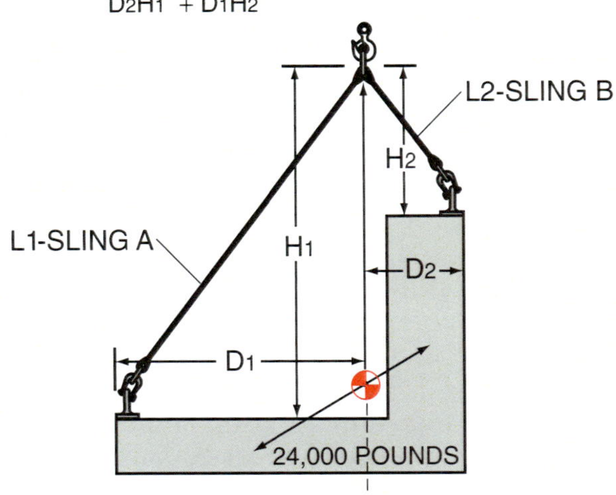

ASYMMETRICAL SLING EXAMPLE

WEIGHT = 24,000 POUNDS

$L_1 = 29.7$ (SLING A) $L_2 = 12$ (SLING B)

$D_1 = 19$ $D_2 = 5$

$H_1 = 22.8$ $H_2 = 10.9$

$$\text{SLING A} = \frac{\text{LOAD} \times D_2L_1}{D_2H_1 + D_1H_2}$$

$$\text{SLING A} = \frac{24{,}000 \times 5 \times 29.7}{(5 \times 22.8) + (19 \times 10.9)}$$

$$\text{SLING A} = \frac{3{,}564{,}000}{321.1}$$

SLING A = 11,099 POUNDS

$$\text{SLING B} = \frac{\text{LOAD} \times D_1L_2}{D_2H_1 + D_1H_2}$$

$$\text{SLING B} = \frac{24{,}000 \times 19 \times 12}{(5 \times 22.8) + (19 \times 10.9)}$$

$$\text{SLING B} = \frac{5{,}472{,}000}{321.1}$$

SLING B = 17,041 POUNDS

NOTE: IRREGULAR LOADS CAN CREATE UNEXPECTED STRESSES. ALWAYS USE A MINIMUM 6:1 SAFETY FACTOR FOR THE RIGGING COMPONENTS.

81201-11_F07.EPS

Figure 7 Asymmetrical bridle sling stress arrangement and formula.

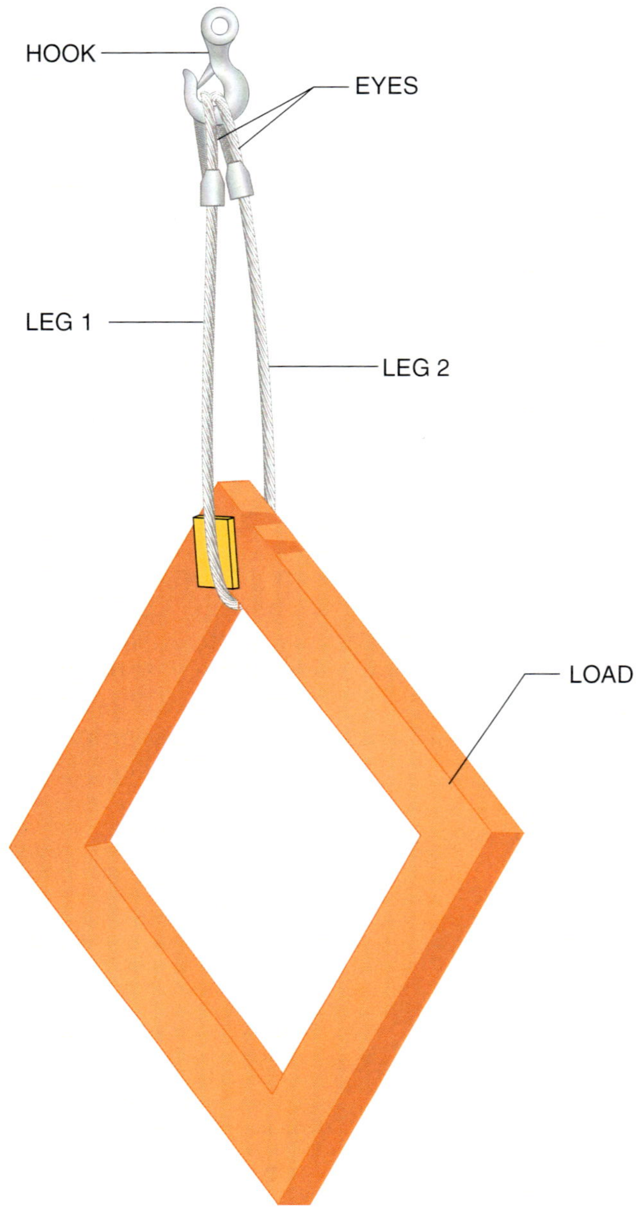

Figure 8 Basket hitch.

- *Single-leg choker hitch* – For leg angles of 45 degrees or less inside the choke, multiply the SWL of the single vertical hitch by ¾. Refer to *Figure 9A*. This can be checked by using the following formula:

  SWL of a single vertical hitch × (A ÷ B)

- *Double-leg choker hitch* – When using more than one sling with a choker hitch, use the single leg choker hitch formula combined with the calculations for bridle hitches. Refer to *Figure 9B*. The following formula applies:

  SWL of a single vertical hitch × (A ÷ B) × (H ÷ L) × 2

- *Double-wrapped choker hitch* – The values for the double-wrapped choker hitch are the same as those for the single choker hitch or the double choker hitch, depending on the configuration.
- *Rendering the choker hitch* – The capacity of a choker hitch is affected by the angle at the point of choke (*Figure 10*). If the choker is forced down to tighten the eye, it must be derated. This commonly happens when turning loads or erecting oddly shaped loads. The formula is:

  SWL of a single vertical hitch × (A ÷ B) × (H ÷ L) × 2 × percent for angle of choke

It is easy to overload a choker hitch if the many capacity reducing factors are not considered for every application.

4.1.5 Rigging for a Top and Trailing Lift

In some instances, a top and trailing lift, usually considered a blind lift, is used to raise and set certain transmission structures. This especially true if a fully assembled tall and heavy monopole steel or concrete transmission structure is to be raised and set. A high-lift mobile crane is used for the top lift and a smaller mobile crane or other mobile lifting device is used for the trailing lift. The top lift crane and its rigging is always selected to safely support the entire weight of the pole when in a vertical position. The strength of a trailing lift device and rigging is selected to initially support the lower end of the pole in the horizontal position based on the CG of the pole. This is because the weight trails off as the pole is raised to nearly the vertical position.

During the lift, the trailing lift device raises the base of the pole off the ground and holds it in that position while the top lift crane raises the top of the pole. As the pole top is being raised, the trailing lift device is moved as required toward the top lift crane to keep the top of pole as close as possible to a position directly under the load-line sheave of the crane. When the pole is in a nearly vertical position, the trailing lift rigging is lowered and allowed to go slack before the pole reaches the vertical position. This is done so the pole doesn't contact the trailing lift device and to allow final setting of the pole. A top and trailing lift requires careful coordination and communication between skilled riggers and the crane/device operators because the position of the pole top in relation to the top lift crane load-line sheave is critical to prevent possible contact with or destabilization of the crane.

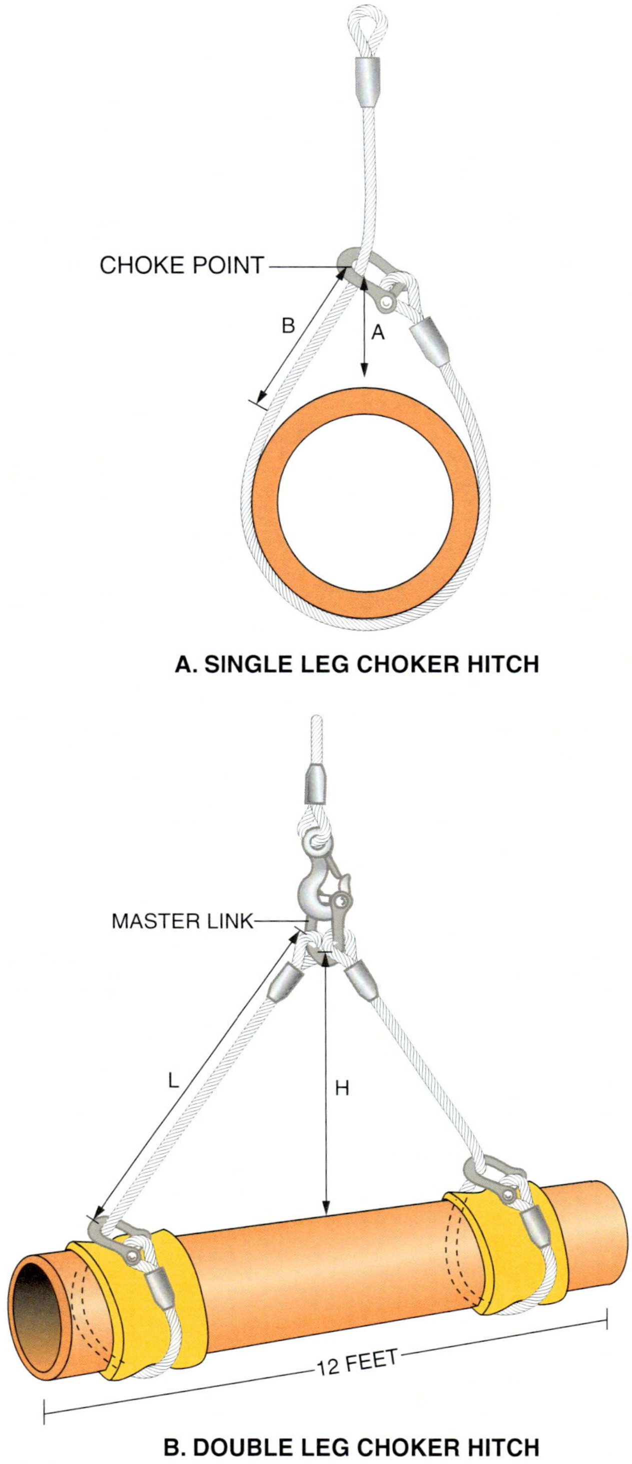

Figure 9 Choker hitch arrangements.

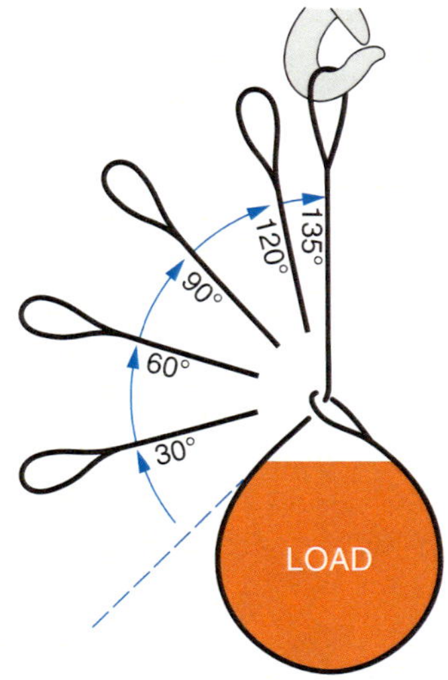

Angle Of Choke (in Degrees)	Rated Capacity IWRC & F.C. Rope Percent*
121–135	95
91–121	82
61–90	71
31–60	60
0–30	48

*Percent of sling capacity in choker hitch.

81201-11_F10.EPS

Figure 10 Choker hitch derating factors.

4.2.0 Standard and Custom Spreader/ Lifting Beams and Slings

There are benefits of using spreader/lifting beams for load handling. A two-legged sling connected to a load exerts a horizontal force across the load. Sling angles of 30 degrees to the load can create a force across the load that is twice the weight of the load. This force can cause damage by compression or buckling of the load. Because of this, sling angles of less than 45 degrees are generally not used. Spreader/lifting beams can completely eliminate the horizontal force by making the sling legs between the beam and the load vertical. They can also contribute to easier handling of the load.

Both fixed-position and adjustable-position spreader/lifting beams are available. However, the adjustable-position spreader/lifting beams shown in *Figure 11* are more versatile. Each beam has selectable or adjustable positions along both the top and bottom of the beam. If used as spreader beams with slings from a crane hook attached to the top of the beam and with vertical slings from the beam to the load, considerable vertical space is required between the crane hook and the load. When used as lifting beams, the crane hook can be connected to an appropriate position on the top of the beam and the only slings required are the vertical ones to the load thus reducing the vertical distance from the hook to the load. Sling strength calculations and asymmetrical load calculations are the same as described in the previous section. If a lifting beam is used, all sling angle derating factors are eliminated except for choker hitches and D/d ratios.

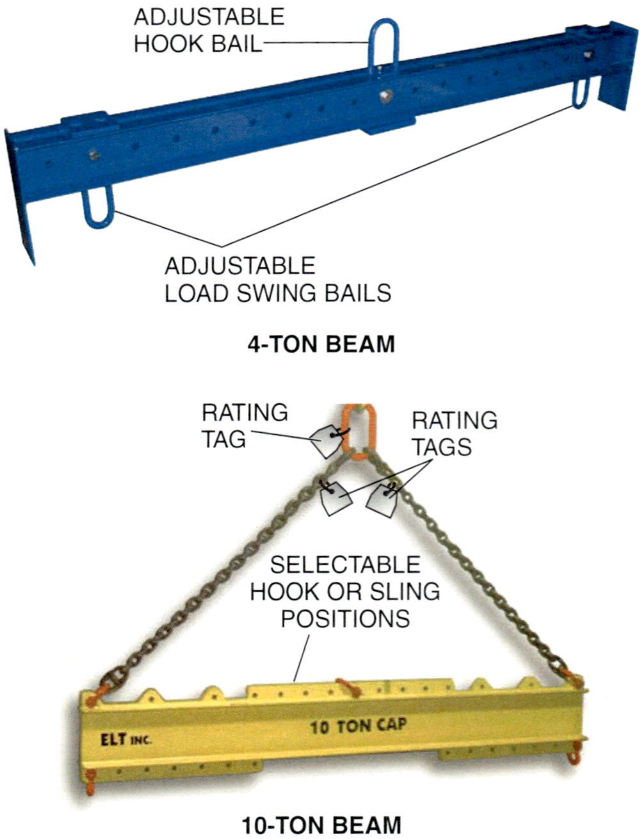

Figure 11 Adjustable-position spreader/lifting beams.

Examples of custom transmission structure rigging are shown in *Figures* 12 and 13. Field-assembled insulated lifting beam rigging is shown in *Figure 12*. It is used to replace a defective horizontal string insulator with a new one being assembled on the ground. The replacement is done in a live voltage (hot) situation. *Figure 13* shows customized sling rigging used during a helicopter lift to place a steel transmission monopole at an inaccessible site.

Figure 12 Field-assembled insulated lifting beam rigging.

Figure 13 Helicopter setting of a steel transmission pole.

5.0.0 Crane Stability

During lifts, crane stability depends on several factors. These factors include the quadrant of operation of the crane, forward and backward stability, noncentered lifts, and environmental factors. As a crane swings to various quadrants, the distance that the load is positioned from the crane's tipping axis changes dramatically. This is because cranes are typically built to be longer than they are wide to allow them to fit on a roadway or trailer for transportation from job to job. This rectangular footprint tends to result in more stability of the crane from front to rear than from side to side.

5.1.0 Quadrants of Operation

Manufacturers define the operating quadrants for each of the cranes that they produce. *Figure 14* shows one example. Along with the definition of the quadrants, the manufacturer also identifies the authorized quadrants of operation and specifies any limitations while operating in those areas. The quadrants of operation are defined by referencing the boom position in relation to its position over the carrier body and the tipping fulcrum (tipping axis) within each quadrant. How the tipping axis affects crane stability is covered in the next section. The quadrants are as follows:

- *Over-the-front* – An area located between the front outriggers or crawlers when facing in the forward direction of travel.
- *Over-the-rear* – The area between the rear outriggers or crawlers when facing the rearward direction of travel.
- *Over-the-side* – A position located between the left- or right-side outriggers or over the crawler.
- *360-degree* – The circular swing area that would encompass all of the previously discussed quadrant.

Rough-terrain cranes and truck-mounted mobile cranes operating on rubber tires have only very limited load capabilities in the over-the side quadrants without the outriggers deployed because the tipping axes are close to the crane. As can be seen in *Figure 14*, when the outriggers are deployed, the extent of the side quadrant area is reduced and the front and rear quadrant areas are expanded. The load capability in the side quadrants is increased because the tipping axes are shifted away from the crane. Without outriggers, rough-terrain cranes are operated only in very narrow front or rear quadrants. Truck-mounted mobile cranes operated on rubber are only used in a very narrow rear quadrant. Even when outriggers are used, truck-mounted mobile cranes are typically operated only in the side or rear quadrants. The front quadrant has limited use because of the boom extension and/or low boom angles required to clear the front of the truck cab and engine.

5.2.0 Centers of Gravity and Crane Stability

The movements of the various centers of gravity in relation to the tipping fulcrum (tipping axis) while operating a crane affect the crane's stability. The most important things to understand are the crane's combined center of gravity and the load's center of gravity. The center of gravity of the crane is a combination of the upperworks (superstructure), base section, and counterweight centers of gravity. Changes that cause movement in a crane's center of gravity are reconfiguring the counterweights, rotating the upperworks, extending or retracting the outriggers, and extending or retracting the crawlers (see *Figure 14*). The crane's tipping axis is typically closer to the upperworks in the side-to-side quadrants. The movements that affect the load's center of gravity in relation to the tipping axis are extending, retracting, raising, and lowering the boom.

The distribution of the forces on either side of the tipping axis determines the stability of the crane. As the crane's combined center of gravity moves toward the tipping axis, the calculated leverage applied by the crane is reduced. This situation also occurs if the load is moved further from the tipping axis. Moving the load further from the tipping axis increases the leverage applied by the load. Both of these conditions cause the leverage forces to equalize, producing an unstable crane condition. The point of equalization of the leverage forces is expressed by the following equation where:

A (crane CG weight) times B (distance from the tipping axis) = C (load CG weight) times D (distance from the tipping axis) or

$$A \times B = C \times D$$

If any three of the four variables are known, such as crane CG weight, crane distance, and load CG weight, the equalization of leverage equation can be solved for the unknown variable D (load distance) as follows:

$$(A \times B)/C = D$$

It is important to remember that at the point where the leverages are equal, the crane is unstable. The crane leverage must always exceed the load leverage. The following mathematical examples elaborate on this relationship.

Example One

Assume a load is being lifted over the front of a crane at a certain boom angle and boom length. The center of gravity (CG) weights of the load and the crane and their distances from the tipping axis are shown in *Figure 15*. The forces applied on each side of the tipping axis are determined by multiplying the (CG) weight of the object by its distance from the tipping axis.

Leverage of the crane:

4,000 lbs (crane weight) ×
8 ft (distance of crane CG from tipping axis)
= 32,000 ft-lbs (leverage of crane)

Leverage of the load (load moment):

500 lbs (load weight) ×
20 ft (distance of load CG from tipping axis)
= 10,000 ft-lbs (leverage of load)

The crane's leverage (32,000 ft-lbs) is much greater than the load moment (10,000 ft-lbs). The crane will be stable.

Example Two

Assume that the crane has been rotated to lift the load from the over-the-side position. The crane weight, load weight, boom angle, and boom

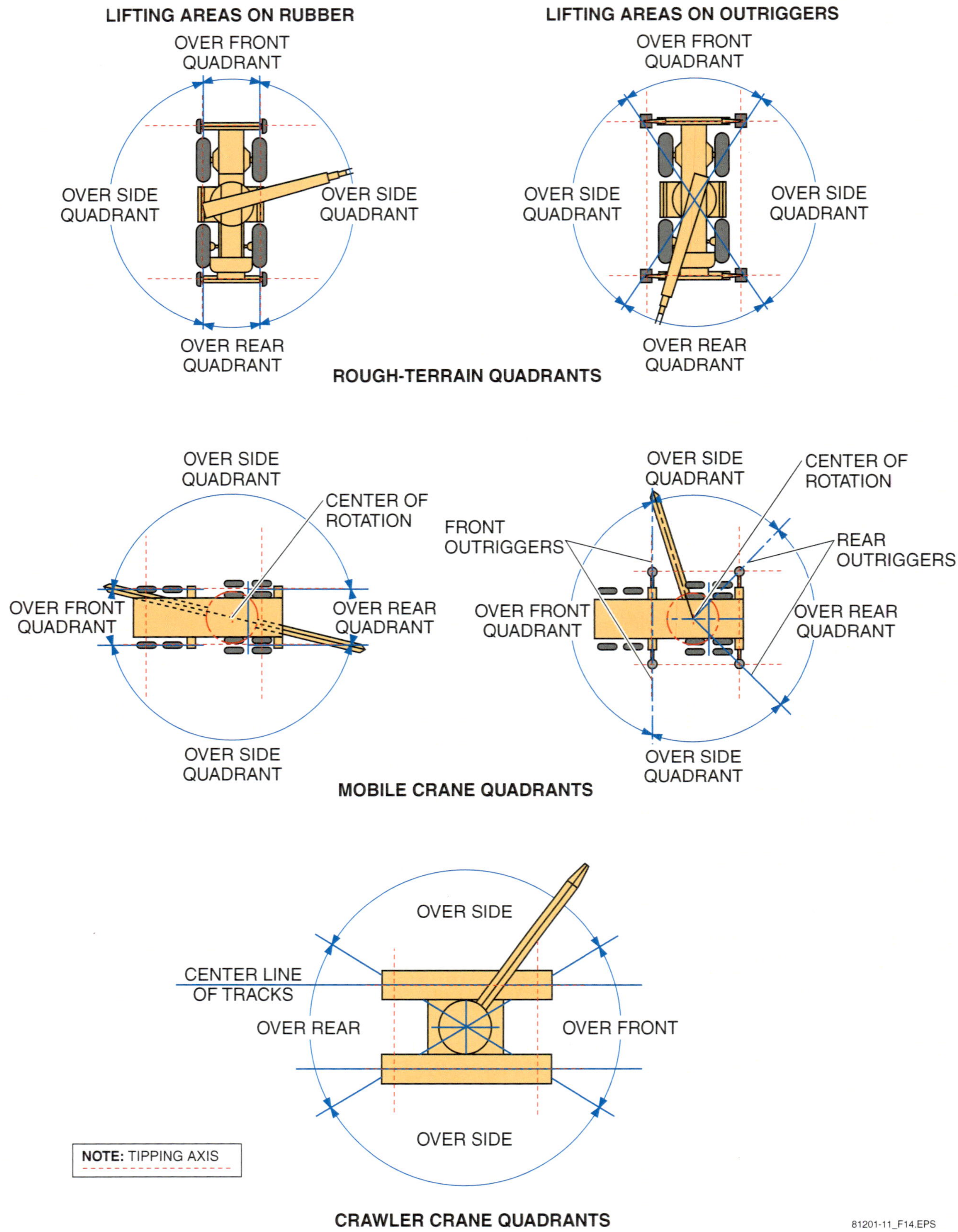

Figure 14 Operational crane quadrants with tipping axis.

length remain the same, as shown in *Figure 16*, but the distances from the tipping axis have changed.

The forces applied on each side of the tipping axis are determined by multiplying the weight of the object by its distance from the tipping axis.

Leverage of the crane:

4,000 lbs (crane weight) ×
3 ft (distance of crane CG from tipping axis)
= 12,000 ft-lbs (leverage of crane)

Leverage of the load (load moment):

500 lbs (load weight) ×
24 ft (distance of load CG from tipping axis)
= 12,000 ft-lbs (leverage of load)

The crane's leverage (12,000 ft-lbs) is equal to the load moment (12,000 ft-lbs). This situation should not be attempted, as the crane will be very unstable. Also note that the tipping axis on the crane moved as the crane was rotated. The new tipping axis remained between the load and the crane's center of rotation. Changing the crane's boom length and raising or lowering the boom (increasing or decreasing the boom angle) as well as moving the boom from one operating quadrant to another can have a dramatic effect on crane stability. When an operator takes action that reduces the stabilizing force on the crane, it can result in a crane tipping accident. Certified crane operators using the crane's load charts are responsible for determining the load capacity of the crane in its operating quadrants for the loads to be lifted.

A crane's load charts dictate the maximum gross loads that can be lifted under ideal conditions in the crane's quadrants based on a specified boom length and counterweight at various boom angles and load radiuses. The tipping axes of the crane are factored into the gross load calculations for the crane based on the operational conditions listed in the load charts.

Figure 17 defines boom length, boom angle, and operating radius for various configurations of truck-mounted mobile cranes. Some manufacturers may also specify boom elevation. *Figure 18* shows a portion of a load chart for a truck-mounted hydraulic boom with its outriggers fully deployed and a specific counterweight and boom configuration. The shaded areas show load capacity that is primarily limited by boom strength. The unshaded areas show load capacity limited only by crane stability. The load capacity in both areas have been derated to 75 percent of the amount that would cause tipping in accordance with *ANSI/SAE J765*. In no case can capacities be interpolated when boom lengths are between the stated lengths. Only the capacities for the next longer boom length can be used. The loads to be lifted must include the weight of the rigging and any accessories or other items as specified by the crane manufacturer.

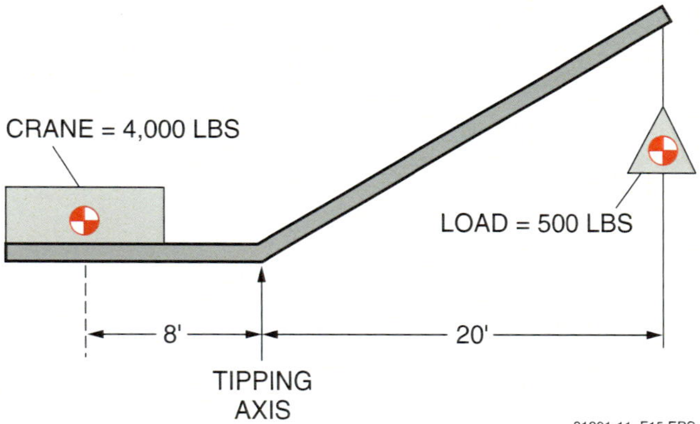

Figure 15 Stability Example One.

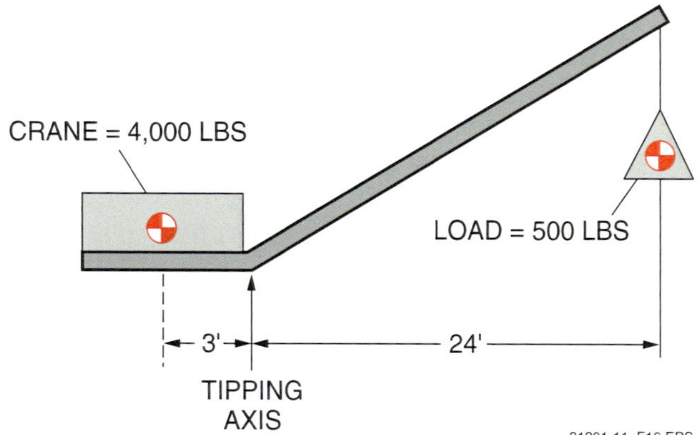

Figure 16 Stability Example Two.

6.0.0 Personnel Platform Requirements and Usage

In compliance with *OSHA 29 CFR 1926.550*, platform requirements are broken down into four categories: design criteria [*29 CFR 1926.550(g)(4)(i)*]; platform specifications [*29 CFR 1926.550(g)(4)(ii)*]; platform loading [*29 CFR 1926.550(g)(4)(iii)*]; and rigging [*29 CFR 1926.550(g)(4)(iv)*]. These requirements are detailed in the following sections.

6.1.0 Design Criteria

The suspended personnel platform, also called a manbasket or workbasket (*Figure 19*), shall be designed by a qualified engineer or a competent person qualified in structural design.

The suspension system shall be designed to minimize tipping of the platform due to movement of the employees occupying the platform.

The personnel platform itself, except the guardrail system and personal fall-arrest system an-

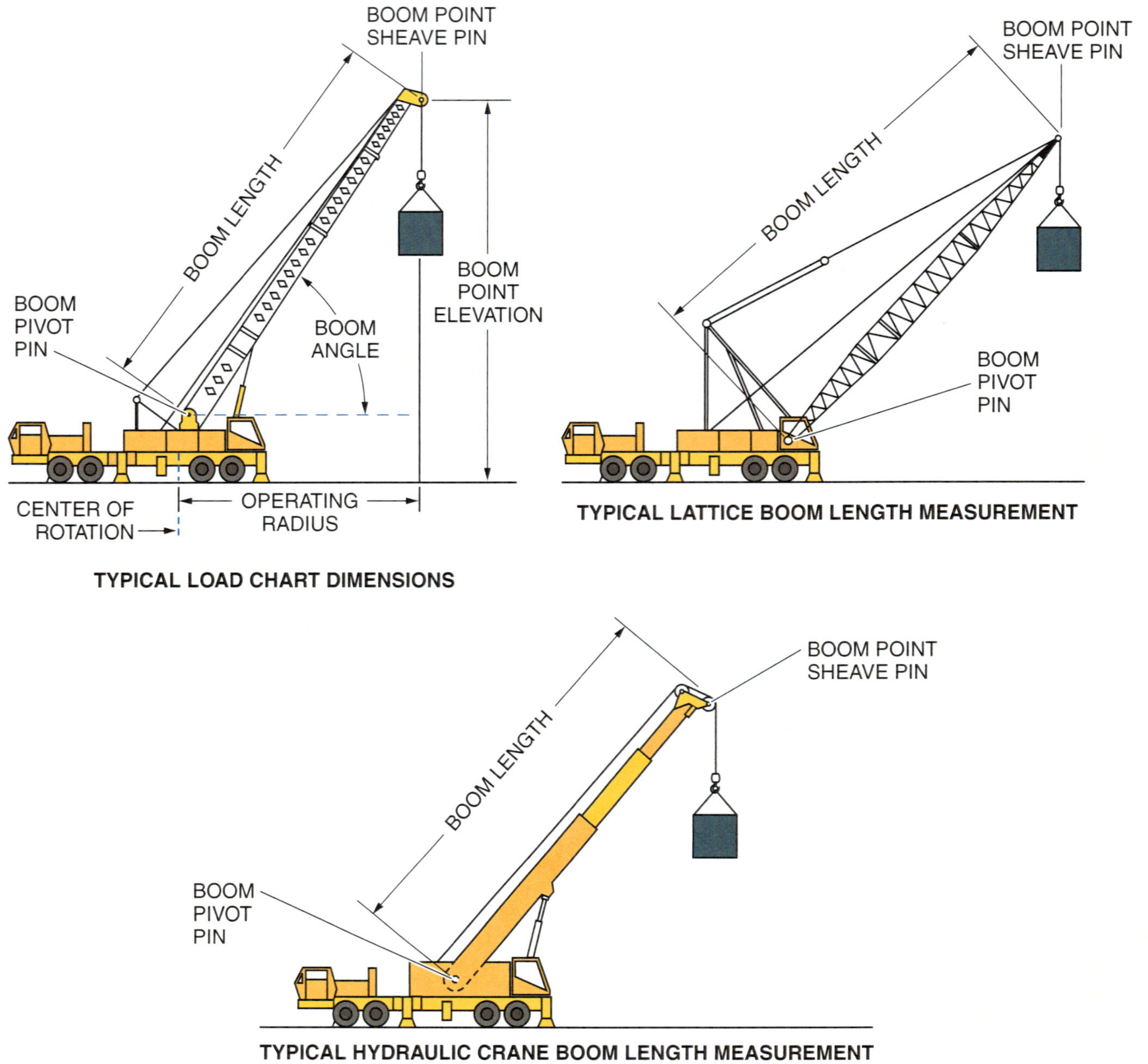

81201-11_F17.EPS

Figure 17 Boom length, boom angle, load operating radius, and boom elevation definitions.

chorages, shall be capable of supporting, without failure, its own weight and at least five times the maximum intended load.

> **NOTE**
> In the following sections taken from OSHA and ASME standards, a suspended personnel platform may be called a platform or a personnel platform.

6.2.0 Platform Specifications

Each personnel platform shall be equipped with a guardrail system and shall be enclosed at least from the toeboard to mid-rail with either solid construction or expanded metal having openings no greater than ½ inch (1.27 cm). Other platform specifications are as follows:

- A grab rail shall be installed inside the entire perimeter of the personnel platform.
- Access gates, if installed, shall not swing outward during hoisting.
- Access gates, including sliding or folding gate, shall be equipped with a restraining device to prevent accidental opening.
- Head room shall be provided which allows employees to stand upright in the platform.
- In addition to the use of hard hats, employees shall be guarded by overhead protection on the personnel platform when employees are exposed to falling objects.

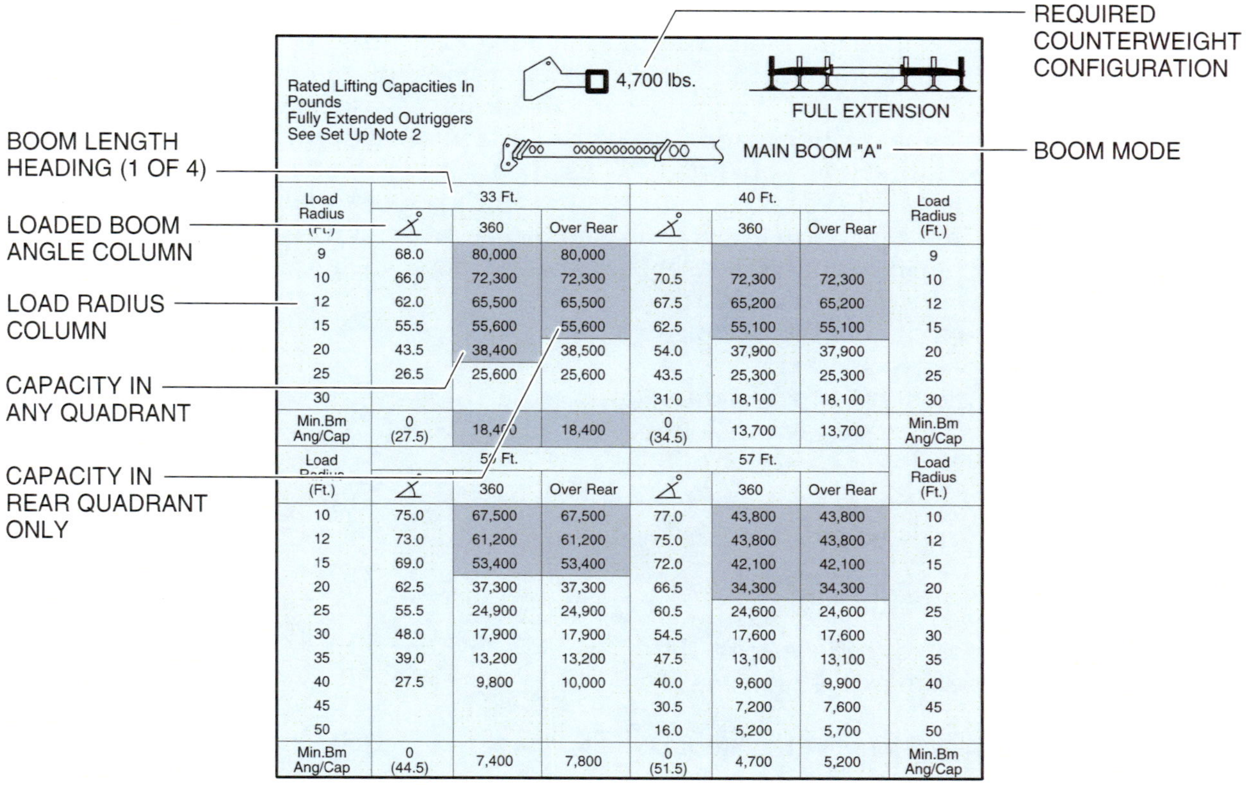

Load Radius (Ft.)	33 Ft. Loaded Boom Angle	33 Ft. 360	33 Ft. Over Rear	40 Ft. Loaded Boom Angle	40 Ft. 360	40 Ft. Over Rear	Load Radius (Ft.)
9	68.0	80,000	80,000				9
10	66.0	72,300	72,300	70.5	72,300	72,300	10
12	62.0	65,500	65,500	67.5	65,200	65,200	12
15	55.5	55,600	55,600	62.5	55,100	55,100	15
20	43.5	38,400	38,500	54.0	37,900	37,900	20
25	26.5	25,600	25,600	43.5	25,300	25,300	25
30				31.0	18,100	18,100	30
Min.Bm Ang/Cap	0 (27.5)	18,400	18,400	0 (34.5)	13,700	13,700	Min.Bm Ang/Cap
Load Radius (Ft.)	**50 Ft. Loaded Boom Angle**	**50 Ft. 360**	**50 Ft. Over Rear**	**57 Ft. Loaded Boom Angle**	**57 Ft. 360**	**57 Ft. Over Rear**	**Load Radius (Ft.)**
10	75.0	67,500	67,500	77.0	43,800	43,800	10
12	73.0	61,200	61,200	75.0	43,800	43,800	12
15	69.0	53,400	53,400	72.0	42,100	42,100	15
20	62.5	37,300	37,300	66.5	34,300	34,300	20
25	55.5	24,900	24,900	60.5	24,600	24,600	25
30	48.0	17,900	17,900	54.5	17,600	17,600	30
35	39.0	13,200	13,200	47.5	13,100	13,100	35
40	27.5	9,800	10,000	40.0	9,600	9,900	40
45				30.5	7,200	7,600	45
50				16.0	5,200	5,700	50
Min.Bm Ang/Cap	0 (44.5)	7,400	7,800	0 (51.5)	4,700	5,200	Min.Bm Ang/Cap

Note: Refer to page 5 for (Capacity Deductions for Auxiliary Load Handling Equipment).

(Boom angle symbol) Loaded Boom Angle in Degrees

( ) Reference radius for minimum boom angle capacities (shown in parenthesis) are in feet.

81201-11_F18.EPS

Figure 18 Hydraulic boom load/capacity chart.

Note: Also available with optional overhead protection if required

81201-11_F19.EPS

Figure 19 Suspended personnel platform.

- All rough edges exposed to contact by employees shall be surfaced or smoothed in order to prevent injury to employees from punctures or lacerations.
- All welding of the personnel platform and its components shall be performed by a qualified welder familiar with the weld grades, types, and material specified in the platform design.
- The personnel platform shall be conspicuously posted with a plate or other permanent marking which indicates the weight of the platform and its rated load capacity or maximum intended load. The maximum intended load includes personnel, tools, accessory equipment, and work materials.

6.3.0 Personnel Platform Loading

The personnel platform shall not be loaded in excess of its rated load capacity. When a personnel platform does not have a rated load capacity, the platform shall not be loaded in excess of its maximum intended load.

The number of employees, along with material, occupying the personnel platform shall not exceed the limit established for the platform and the rated load capacity or the maximum intended load.

Personnel platforms shall be used only for employees, their tools, and the materials necessary to do their work, and shall not be used to hoist only materials or tools when not hoisting personnel.

Materials and tools for use during a personnel lift shall be secured to prevent displacement. These items shall be evenly distributed within the confines of the platform while the platform is suspended.

6.4.0 Rigging

When a four-leg wire rope or chain bridle sling is used to connect the personnel platform to the load line, each bridle leg shall be connected to a master link or shackle in such a manner as to ensure that the load is evenly divided among the bridle legs (*Figures 20* and *21*).

Hooks on headache ball assemblies, lower load blocks, or other attachment assemblies shall be of a type that can be closed and locked, eliminating the hook throat opening. Alternatively, an alloy anchor type shackle with a bolt, nut, and retaining pin may be used.

Wire rope, shackles, rings, master links, and other rigging hardware must be capable of supporting, without failure, at least five times the maximum intended load applied or transmitted to that component. Where rotation-resistant rope is used, the slings shall be capable of supporting without failure at least ten times the maximum intended load. All eyes in wire rope slings shall be fabricated with thimbles.

Bridles and associated rigging for attaching the personnel platform to the hoist line shall be used only for the platform and the necessary employees, their tools, and the materials necessary to do their work, and they shall not be used for any other purpose when not hoisting personnel.

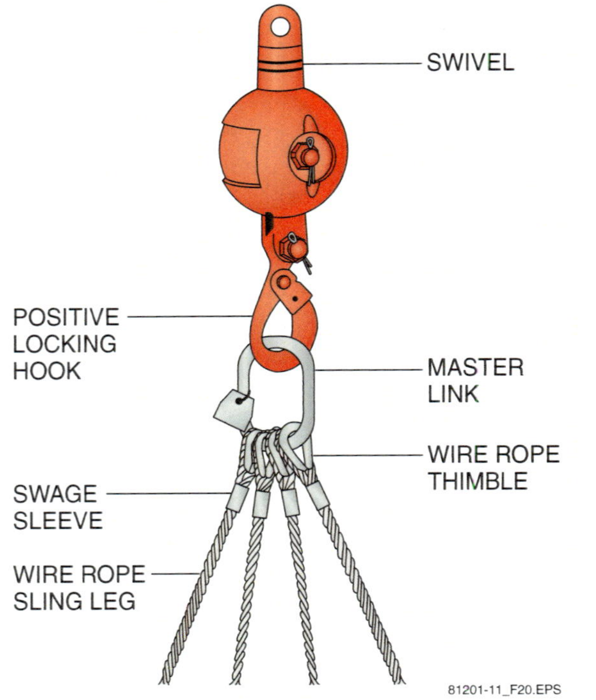

Figure 20 Bridle sling using a master hook.

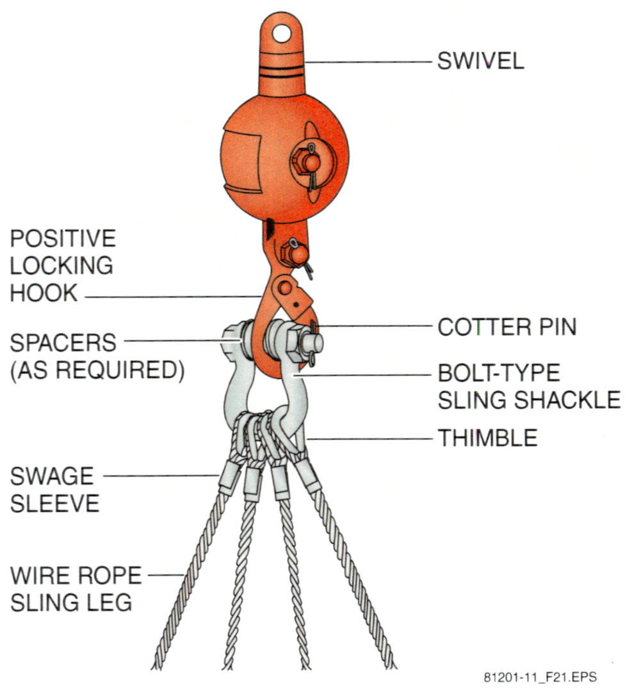

Figure 21 Bridle sling using a shackle.

6.5.0 Crane and Operational Requirements

According to *OSHA 29 CFR 1926.550(g)*, crane and operational requirements are broken down into operational criteria [*29 CFR 1926.550(g)(3)(i)*]; instruments and components [*29 CFR 1926.550(g)(3)(ii)*]; work practices [*29 CFR T1926.550(g)(6)*]; and traveling [*29 CFR 1926.550(g)(7)*]. These requirements are given in detail in the following sections.

6.5.1 Operational Criteria

Hoisting of the personnel platform shall be performed in a slow, controlled, cautious manner with no sudden movements of the crane or derrick or the platform.

Load lines shall be capable of supporting, without failure, at least seven times the maximum intended load. Where rotation-resistant rope is used, the lines shall be capable of supporting, without failure, at least ten times the maximum intended load. The required design factor is achieved by taking the current safety factor of 3.5, which is required under *OSHA 1926.550(b)(2)*, and applying the 50 percent derating of the crane capacity which is required by *OSHA 1926.550(g)(3)(i)(f)*.

Load and boom hoist drum brakes, swing brakes, and locking devices, such as pawls or dogs, shall be engaged when the occupied personnel platform is in a stationary position.

The crane shall be uniformly level within one percent of level grade and located on firm footing.

Cranes equipped with outriggers shall have them all fully deployed following manufacturer's specifications, insofar as applicable, when hoisting personnel.

The total weight of the loaded personnel platform and related rigging shall not exceed 50 percent of the rated capacity for the radius and configuration of the crane or derrick.

The use of machines having booms in which lowering is controlled by a brake without aid from other devices which slow the lowering speeds is prohibited.

6.5.2 Instruments and Components

Cranes and derricks with variable-angle booms shall be equipped with a boom angle indicator, readily visible to the operator.

Cranes with telescoping booms shall be equipped with a device to indicate clearly to the operator, at all times, the boom's extended length, or an accurate determination of the load radius to be used during the lift shall be made prior to hoisting personnel.

A positive-acting device shall be used which prevents contact between the load block or headache ball and the boom tip (anti-two-blocking device), or a system shall be used which deactivates the hoisting action before damage occurs in the event of a two-blocking situation (two-block damage prevention feature).

The load-line hoist drum shall have a system or device on the power train, other than the load hoist brake, which regulates the lowering rate of speed of the hoist mechanism (controlled load lowering). Free fall is prohibited.

6.5.3 Work Practices

The following work practices must be observed when using a suspended personnel platform. As defined in *29 CFR 1926.550(g)(3)(i)* and *OSHA 1926.106*, these requirements are detailed as follows:

- Personnel shall keep all parts of the body inside the platform during raising, lowering, and positioning. This provision does not apply to an occupant of the platform performing the duties of a signal person.
- Before workers exit or enter a hoisted personnel platform that is not landed, the platform shall be secured to the structure where the work is to be performed, unless securing to the structure creates an unsafe situation.
- Tag lines shall be used unless their use creates an unsafe situation.
- The crane or derrick operator shall remain at the controls at all times when the crane engine is running and the platform is occupied.
- Hoisting of personnel shall be promptly discontinued upon indication of any dangerous weather conditions or other impending danger.
- Personnel being hoisted shall remain in continuous sight of, and in direct communication with, the operator or signal person. In those situations where direct visual contact with the operator is not possible, and the use of a signal person would create a great hazard for the person, direct communication alone (such as by radio) may be used.
- Except over water, workers occupying the personnel platform shall use a body belt/harness system with a lanyard appropriately attached to the lower load block or headache ball, or to a structural member within the personnel platform capable of supporting a fall impact for personnel using the anchorage. When working over water, the requirements of *OSHA 1926.106* shall apply.

- No lifts shall be made on another of the crane's or derrick's load lines while personnel are suspended on a platform.
- Employees working over or near water where the danger of drowning exists shall be provided with US Coast Guard-approved life jackets or buoyant work vests.
- Prior to and after each use, the buoyant work vests or life preservers shall be inspected for defects which would alter their strength or buoyancy. Defective units shall not be used.
- Ring buoys with at least 90 feet of line shall be provided and readily available for emergency rescue operations. Distance between ring buoys shall not exceed 200 feet.
- At least one lifesaving skiff shall be immediately available at locations where personnel are working over (or adjacent to) water.

6.5.4 Traveling

Hoisting personnel while the crane is traveling is prohibited, except for portal, tower, and locomotive cranes or where the supervisor demonstrates that there is no less hazardous way to perform the work.

Under any circumstances where a crane would travel while hoisting personnel, the employer or supervisor shall implement the following procedures to safeguard personnel:

- Crane travel shall be restricted to a fixed track or runway.
- Travel shall be limited to the load radius of the boom used during the lift.
- The boom must be parallel to the direction of travel.
- A complete trial run shall be performed to test the route of travel before employees are allowed to occupy the platform.
- If travel is done with a rubber-tired carrier, the condition and air pressure of the tires shall be checked. The chart capacity for personnel lifts on rubber shall be used for application of the 50-percent reduction of rated capacity.
- Crane travel speed must be as slow as possible and sudden stops and starts should be avoided.

6.6.0 Personnel Platform Inspection

According to *OSHA 29 CFR 1925.550(g)* guidelines, a visual inspection of the crane or derrick, rigging, personnel platform, and the crane or derrick base support or ground shall be conducted by a competent person immediately after the trial lift with the platform suspended several inches. This inspection is made to determine whether the testing has exposed any defect or produced any adverse effect upon any component or structure. Personnel shall not be hoisted unless the following conditions are determined to exist:

- Hoist ropes shall be free of kinks.
- Multiple-part lines shall not be twisted around each other.
- The primary attachment shall be centered over the platform.
- Fall protection equipment must be free of damage and defects.
- The hoisting system shall be inspected if the load rope is slack to ensure that all ropes are properly seated on drums and in sheaves.

Any defects found during inspections which create a safety hazard shall be corrected before hoisting personnel.

6.6.1 ASME-Prescribed Personnel Platform Inspection

ASME B30.23 guidelines break down inspections into three categories: initial, frequent, and periodic. Prior to use at each new job site, all platforms shall be inspected by a qualified person to verify compliance with the provisions of *ASME B30.23*. This is the initial inspection.

During a frequent inspection, the platform, suspension system, and attachment points shall be inspected by a designated person. This inspection is made to identify conditions that have been specifically indicated by the platform manufacturer or a qualified person designated by the employer as potentially creating a hazardous operating condition. The frequent inspection shall address at least the following items:

- Markings
- Structure
- Attachment mechanisms
- Special purpose items

Periodic inspections of the same items are to be conducted at least once every 12 months. Platforms that have been out of service for 12 or more consecutive months shall receive a periodic inspection prior to use.

Dated records of frequent inspections must be made and kept by the platform user for the duration of the personnel lift operation. Dated records of periodic inspections must be made by the platform owner. The last inspection shall be kept with the platform and be available for review.

6.7.0 Trial Lift

The requirements for conducting a trial lift under *OSHA 29 CFR 1926.550(g)(5)* are as follows:

- A trial lift with the unoccupied personnel platform loaded at least to the anticipated lift weight shall be made from ground level or any other location where personnel enter the platform, to each location at which the personnel platform is to be hoisted and positioned. This trial lift shall be performed immediately prior to placing personnel on the platform. The operator shall determine that all systems, controls, and safety devices are activated and functioning properly; that no interferences exist; and that all configurations necessary to reach those work locations allow the operator to remain under the 50-percent limit of the hoist's rated capacity. A single trial lift may be performed at one time for all locations that are to be reached from a single setup position.
- The trial lift shall be repeated prior to hoisting personnel whenever the crane or derrick is moved and set up in a new location or returned to a previously used location. Additionally, the trial lift shall be repeated when the lift route is changed, unless the operator determines that the route change is not significant.
- At each job site, prior to hoisting personnel, the platform and rigging shall be proof-tested to 125 percent of the platform's rated capacity. This is accomplished by holding the platform in a suspended position for five minutes with the test load evenly distributed on the platform. Proof testing can be accomplished concurrently with the trial lift. After proof testing the platform, a competent person shall inspect the platform and rigging. If any deficiencies are found, they shall be corrected and another test conducted. Personnel hoisting shall not take place until the proof testing requirements are satisfied.

SUMMARY

Specialized knowledge is needed to plan and implement rigging operations. Many factors can affect a lifting operation, and each factor can cause a load to act in a different manner. It is up to the crane operator and rigging specialist to understand how to prevent conditions that might cause an unstable load or an unstable crane.

Although engineers may design a lift, the rigging specialist plays an important role in the planning and in carrying out the lift. At every step in the process, the rigging specialist must ensure that the operation is carried out safely and that the correct rigging equipment and methods are being used. Personnel involved in rigging must also be aware of the rules and regulations for the use of personnel platforms.

Review Questions

1. A blind lift is a lift where the crane operator cannot see the load movement throughout the entire lift.
   a. True
   b. False
2. The formula for determining the SWL of a bridle hitch is _____.
   a. SWL = W(L ÷ H)No. of legs
   b. SWL = W(H ÷ L)No. of legs
   c. SWL = L(W ÷ H)No. of legs
   d. SWL = H(L ÷ W)No. of legs
3. The SWL of a three-leg bridle sling for a symmetrical load requiring 6-foot legs, a center hook height of 3 feet, and using legs with a vertical-pull SWL of 4,000 lbs is _____.
   a. 2,000 lbs
   b. 4,000 lbs
   c. 6,000 lbs
   d. 8,000 lbs
4. The SWL of a double basket hitch using vertical legs with a vertical-pull SWL of 1,000 lbs is _____.
   a. 1,000 lbs
   b. 2,000 lbs
   c. 3,000 lbs
   d. 4,000 lbs
5. The SWL of a single choker hitch using a vertical-pull sling with an SWL of 1,000 lbs and a leg angle of 45 degrees or less inside the choke is _____.
   a. 250 lbs
   b. 500 lbs
   c. 750 lbs
   d. 1,000 lbs
6. The rectangular footprint of a crane contributes to crane stability in the _____.
   a. front to side position
   b. rear to side position
   c. side to side position
   d. front to rear position
7. With a crane CG weight of 3,000 lbs at 10 feet from the tipping axis and a load CG weight of 1,000 lbs at 20 feet from the tipping axis, will the crane be stable when the load is lifted?
   a. Yes
   b. No
8. Given a crane CG weight of 2,000 lbs at 5 feet from the tipping axis and a load CG weight of 800 lbs, at what distance from the tipping axis does the load make the crane unstable?
   a. 10½ feet
   b. 12½ feet
   c. 14¼ feet
   d. 16¾ feet
9. Given a crane CG weight of 5,000 lbs at 20 feet from the tipping axis and a load CG weight of 2,000 lbs, at what distance from the tipping axis does the load make the crane unstable?
   a. 30 feet
   b. 40 feet
   c. 50 feet
   d. 60 feet
10. A personnel platform must be capable of supporting its own weight and at least how many times the maximum intended load?
    a. two times
    b. three times
    c. four times
    d. five times
11. A grab rail must be installed _____.
    a. on each corner of the platform
    b. inside the perimeter of the platform
    c. one foot above the exterior of the floorboard
    d. at the owner's discretion
12. According to *OSHA 1926.550(g)*, access gates must be equipped with a _____.
    a. restraining device
    b. five-inch handle
    c. ten-inch handle for easy gripping
    d. steel grip device

13. Load lines must be capable of supporting, without failure, at least how many times the maximum intended load?

a. 4 times
b. 5 times
c. 6 times
d. 7 times

14. Controlled load lowering is described as lowering that _____.

a. uses positively locking hooks
b. uses tag lines
c. is regulated by a system or device on the power train, other than the load hoist brake
d. is controlled by a brake without aid from other devices that slow the lowering speed

15. Prior to use, platforms which have been out of service for 12 or more consecutive months must receive a(n) _____.

a. initial inspection
b. frequent inspection
c. periodic inspection
d. annual inspection

Trade Terms Introduced in This Module

Apex: The point of sling attachment to a crane hook.

Blind lift: A lift where the crane operator(s) or a helicopter pilot cannot observe all or part of the load movement.

D/d ratio: The relationship between the diameter of the surface (D) that a wire rope sling is around and the diameter of the sling wire rope (d).

Appendix

Hand Signals

Swing — Extend arm with closed fist, extend index finger. Use appropriate arm for desired direction.
Raise Boom and Lower Load — Extend arm, thumb up, open and close fingers.
Lower Boom and Raise Load — Extend arm, thumb down, open and close fingers.
Travel — Extend arm, palm raised, and motion arm in the direction desired.
Extend Boom — Extend arms in front of body, palms up, fists closed, extend thumbs out to the sides.
Retract Boom — Extend arms in front of body, palms down, fists closed, extend thumbs inward.
Travel Both Tracks — With clenched fists, roll one fist over the other.
Travel One Track — Raise arm, fist clenched, to indicate lock track; roll other fist to travel. Raised hand indicates track to travel.
Stop — Extend arm, palm down, and hold. Move hand and forearm in a horizontal chopping motion.
81201-11_A01A.EPS

81201-11_A01B.EPS

81201-11_A01C.EPS

Additional Resources

This module presents thorough resources for task training. The following resource material suggested for further study.

Bob's Rigging and Crane Handbook, Latest Edition. Leawood, KS: Pellow Engineering Services.

Figure Credits

Photograph courtesy of American Electric Power, Module opener

California Public Utilities Commission, Figure 2

Courtesy of Northern Tool + Equipment and Vestil Manufacturing Corporation, Figure 11 (top photo)

Engineered Lifting Technologies, www.eltlift.com, Figure 11 (bottom photo)

KT Power, Inc., Figure 12

Allteck Line Contractors, Inc. (www.allteck.ca) Vancouver Island Transmission Reinforcement Project, Figure 13

Link-Belt Construction Equipment Co., Figure 18

Lifting Technologies, Inc., Figure 19

Wire Rope Technical Board, Figures 20 (wire illustration) and 21 (wire illustration)

NCCER CURRICULA — USER UPDATE

NCCER makes every effort to keep its textbooks up-to-date and free of technical errors. We appreciate your help in this process. If you find an error, a typographical mistake, or an inaccuracy in NCCER's curricula, please fill out this form (or a photocopy), or complete the online form at **www.nccer.org/olf**. Be sure to include the exact module ID number, page number, a detailed description, and your recommended correction. Your input will be brought to the attention of the Authoring Team. Thank you for your assistance.

Instructors – If you have an idea for improving this textbook, or have found that additional materials were necessary to teach this module effectively, please let us know so that we may present your suggestions to the Authoring Team.

NCCER Product Development and Revision

13614 Progress Blvd., Alachua, FL 32615

Email: curriculum@nccer.org
Online: www.nccer.org/olf

❑ Trainee Guide ❑ AIG ❑ Exam ❑ PowerPoints Other ______________________________

Craft / Level: Copyright Date:

Module ID Number / Title:

Section Number(s):

Description:

Recommended Correction:

Your Name:

Address:

Email: Phone:

81202-11

Transmission Structure Erection

Module Three

Trainees with successful module completions may be eligible for credentialing through NCCER's National Registry. To learn more, go to **www.nccer.org** or contact us at **1.888.622.3720**. Our website has information on the latest product releases and training, as well as online versions of our *Cornerstone* newsletter and Pearson's product catalog.

Your feedback is welcome. You may email your comments to **curriculum@nccer.org,** send general comments and inquiries to **info@nccer.org**, or use the User Update form at the back of this module.

V.1 11/11

81202-11
Transmission Structure Erection

Objectives

When you have completed this module, you will be able to do the following:

1. Describe the different types of transmission structures.
2. Describe transmission structure construction/erection requirements.
3. Explain the different types of transmission structure foundations.
4. Describe different guying methods.
5. Perform a site inspection to determine:
   - Accessibility
   - Hazards
   - Environmental concerns
   - Right-of-way access through private property
6. Using applicable specifications and drawings, frame and erect a wood H-frame tangent transmission structure.
7. Test structure grounds:
   - Ground rods
   - Counterpoise

Performance Tasks

Under the supervision of your instructor, you should be able to do the following:

1. Perform a site inspection to determine:
   - Accessibility
   - Hazards
   - Environmental concerns
   - Right-of-way access through private property
2. Using applicable specifications and drawings, frame and erect a wood H-frame tangent transmission structure.
3. Test structure grounds:
   - Ground rods
   - Counterpoise

Trade Terms

Circuit
Dead-end structure
Deviation/tension structure
Lattice steel tower (LST)
Suspension structure
Transmission structure
Tubular steel pole (TSP)

Industry Recognized Credentials

If you're training through an NCCER-accredited sponsor you may be eligible for credentials from NCCER's Registry. The ID number for this module is 81202-11. Note that this module may have been used in other NCCER curricula and may apply to other level completions. Contact NCCER's Registry at 888.622.3720 or go to www.nccer.org for more information.

Contents

Topics to be presented in this module include:

Figures and Tables

Figures and Tables (continued)

1.0.0 INTRODUCTION

Transmission structures are used to support overhead lines (conductors) for electric power transmission (*Figure 1*). In North America, a transmission structure, typically a **lattice steel tower (LST)** or tubular structure, is commonly called a transmission tower. In some parts of Canada, it may be referred to as a hydro tower. In Europe and other countries, they may be called electricity pylons or ironmen. Transmission structures are used in high-voltage DC and three-phase high-voltage AC transmission networks between and within regional power systems. Within a regional power system, power from three-phase high-voltage transmission lines is supplied to distribution networks for application to the loads. In some cases, DC transmission structures are used as monopolar or bipolar long distance interties between regional power networks. Bipolar DC transmission structures carry both a positive and a negative line for the DC power. Some DC transmission is accomplished using a monopolar (+) line with earth or a buried line as the negative line.

81202-11_F01.EPS

Figure 1 765kV three-phase power transmission structure.

2.0.0 TRANSMISSION STRUCTURES

Transmission structures vary in type, design, and size depending on the terrain and soil conditions over which the lines are strung and the voltage/current being transmitted. The structures typically range in height from 50 to 200 feet (15 to 61 meters) although higher ones are also used.

While structures have been used in the past for voltages as low as 34.5kV, the standard for usage is 69.5kV or more. Today, transmission structures are typically used for voltages ranging from 110kV to 765kV. Voltages from 69.5kV up to 345kV are classified as high-voltage (HV) transmissions and voltages from 345kV to 765kV are classified as extra high-voltage (EHV) transmissions. China is in the process of using newly designed structures for 800kV ultra high-voltage (UHV) transmissions. Ultra high-voltage transmissions of 1,000kV to 2,250kV are currently in the research stages. High-voltage DC transmission is typically from 230kV to 450kV.

Transmission structures are used for three basic functions along the path of high-voltage transmission lines, as shown in *Figure 2*. Some structures may perform a combination of these functions.

Suspension structures, also called tangent structures or towers, are used along relatively straight and level runs of transmission lines. They are also used for long, shallow turns. They use vertical string insulators to support the transmission lines.

Dead-end structures, also called terminal and anchor structures or towers, are normally used at the endpoints of specified lengths of line conduc-

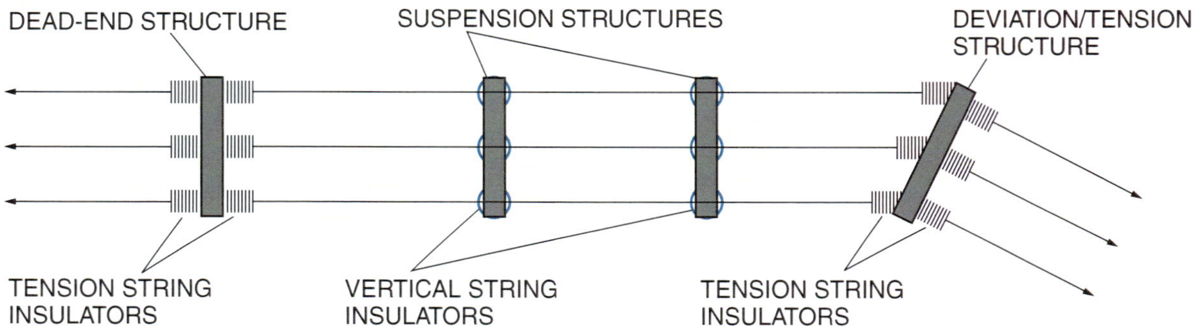

81202-11_F02.EPS

Figure 2 Portion of a path for a high-voltage transmission line.

tors after a specified number of spans between suspension structures. They are also used for a transition from overhead lines to substations or to underground cables. Horizontal tension string insulators are used on dead-end structures to provide the tension for lines between the last and next dead-end structure.

Deviation/tension structures, also called angle structures or towers, are similar to dead-end structures. They are used to sharply change the direction of transmission lines horizontally or on inclines. Tension string insulators are also used on these structures.

The transmission structure's string insulators, which are also called cap and pin insulators, are made of insulator disks that are individually fastened together. Rigid disk insulators are formed as a single unit or as part of a rigid insulator rod. String insulators are normally used as vertical suspension insulators or as horizontal tension insulators. Rigid insulators are sometimes used on deviation/tension and dead-end structures as vertical or horizontal insulators that are rigidly secured to the transmission structure to prevent the conductor drop loop between the horizontal tension insulators from contacting the transmission structure in high winds.

Any set of three transmission lines used on a structure is for three-phase power and is commonly referred to as a **circuit**. Transmission lines for voltages up to 115kV are normally single conductors. However, transmission lines for higher voltages are usually bundled. This means that more than one conductor is used for each phase and is separated by insulated spacers in the spans between structures and at the insulators on the structures. Transmission lines for 110kV up to 230kV may be single conductors or two-conductor bundles. Voltages from 230kV up to 385kV may use two-, three-, or four-bundle conductors and voltages from 385kV up to 765kV may use five- or six-bundle conductors. Bundled conductors are used to reduce corona losses and audible noise from the transmission lines. Due to the skin effect transmission of current at higher voltages, bundled conductors can carry more current than a single conductor for the same amount of weight. However, bundled conductors in excess of two or three are generally not used in areas subject to heavy icing and high wind conditions because of excessive weight and wind loading of the conductor spans between transmission structures. Earth ground lines at the top of the structures are used for lighting protection. Some examples of structures used for transmission lines are described in the following sections.

2.1.0 Lattice Steel Structures

Some of the various lattice steel structures, called lattice steel towers (LSTs), are shown in *Figures 3* through *9*. Transmission structures are normally erected in a vertical position. However, self-supporting four-legged LSTs inclined from the vertical are sometimes used for transmission lines that must descend into and/or ascend from very steep valleys. *Figure 3* shows a four-legged self-supporting two-circuit suspension LST with single conductors and dual vertical string insulators. The structure is a trellis-type LST with three lattice steel arms on each side with ground wires at the top of the structure. Also shown in *Figure 3* is the same type of trellis LST used as a dead-end structure in the same two-circuit transmission path. It uses horizontal string insulators for conductor support. Note that vertical insulators are not used for the drop loops.

Figure 4 shows a four-legged bridge-type suspension LST with a single circuit using two-conductor bundled lines with vertical string insulators. The earth ground lines are on projections at the top of the tower. *Figure 5* shows a similar single-circuit bridge-type suspension LST. It carries 765kV five-conductor bundle transmission lines using V-string insulators for each bundle to prevent lateral movement of the bundles at the structure by the wind.

Figure 6 shows two four-legged, self-supporting bridge-type deviation/tension LSTs with tension string insulators. Each is a single circuit LST. Note that a pair of vertical strings is used for the drop loops of two phases on the prevailing wind side of the structures. The left hand tower uses two-conductor bundles for the lines and the right hand tower uses single-conductor lines. *Figure 7* shows a four-legged, self-supporting bridge-type dead-end LST with two tension string insulators used for each two-conductor bundled line.

Figure 8 shows a two-legged, self-supporting trellis-type suspension LST with two circuits using single-conductor lines with vertical string insulators. *Figure 9* shows a guyed V-type bridged suspension LST for 765kV transmission using V-string insulators and six-conductor bundles for each line. The four substantial guy wires are anchored at four corners a considerable distance from the LST base. While the base provides weight support, the guy wires provide the only lateral and parallel support to keep the LST upright.

TRELLIS SUSPENSION STRUCTURE

TRELLIS DEAD-END STRUCTURE

81202-11_F03.EPS

Figure 3 Four-legged, self-supporting two-circuit trellis-type suspension and dead-end LST.

81202-11_F04.EPS

Figure 4 Four-legged, self-supporting single-circuit bridge-type suspension LST.

2.2.0 Tubular Steel or Concrete/Synthetic Material Structures

Tubular steel, concrete, or synthetic material monopole or H-frame type structures have been increasingly popular for replacement of lattice steel towers and for new transmission line projects. **Tubular steel poles (TSPs)** are simpler to assemble and erect than LSTs, as well as more aesthetically acceptable in urban areas.

Figure 10 shows a pair of 345kV two-circuit self-supporting coated tubular-steel monopole suspension TSPs that are over 130 feet (40 meters) tall. The three sets of davit arms on each side of the poles carry a single circuit using two-bundle conductors per phase supported by V-string insulators. The top davit arms are carrying the earth ground wires. A pair of the same type of TSPs rigged as deviation/tension structures and used for the same transmission path is shown in *Figure 11*. The two-bundle conductors for each phase of the circuits are supported by dual tension string insulators and a vertical string insulator for the drop loops. *Figure 12* shows the self-supporting coated dead-end TSPs used at the beginning of the four transmission circuits covered in the pre-

81202-11_F05.EPS

Figure 5 Top of a four-legged, self-supporting 765kV single-circuit bridge-type suspension LST.

81202-11_F06.EPS

Figure 6 Two four-legged, self-supporting bridge-type deviation/tension LSTs.

81202-11_F07.EPS

Figure 7 Four-legged, self-supporting bridge-type dead-end LST.

81202-11_F08.EPS

Figure 8 Two-legged, self-supporting two-circuit trellis-type suspension LST.

81202-11_F09.EPS

Figure 9 Guyed V-type 765kV single-circuit suspension LST.

vious two figures. Each circuit is carried on a single pole and the two-conductor bundle for each phase of the circuit is supported by dual tension string insulators with a rigid horizontal insulator for the drop loops.

Figure 13 shows a self-supporting galvanized H-frame suspension TSP structure that has been coated. Its cross-arm is carrying a single circuit with two-conductor bundles supported by two vertical insulators and a V-string insulator. TSPs

81202-11_F10.EPS

Figure 10 345kV self-supporting coated two-circuit suspension TSPs.

are also used in the fabrication of X-frame and Y-frame transmission structures. Today, many TSP and LST structures are constructed of weathering steel. Weathering steel is a particular type of alloy steel known as COR-TEN® steel that is currently designated as *ASTM A588* steel. The composition of the elements in the steel allows the formation of a thin stable iron-oxide (rust) layer on the surface that develops and then regenerates continuously as required when exposed to weather. As a result, the steel surface requires no further surface treatment.

2.3.0 Wood Structures

Wood H-frame structures are used in the United States for voltages typically ranging from 69.5kV to 230kV. Today, they are more commonly used for 115kV to 230kV. In some cases, they are used for voltages up to 345kV. Almost all wood H-frames use an X-brace between the poles to provide lateral stability to the structure. *Figure 14* shows a self-supporting wood H-frame suspension (tangent) structure with braced double cross-arms for a single circuit. The double cross-arms are supporting single conductors using vertical string insulators.

81202-11_F11.EPS

Figure 11 345kV self-supporting coated two-circuit deviation/tension TSPs.

Light-duty lower-voltage H-frames may use only a single-side cross-arm or no X-brace. *Figure 15* shows a side-guyed, three-pole single-circuit wood deviation/tension (angle) structure with a dual set of double cross-arms. The single-circuit phase conductors are suspended by single and dual tension string insulators and vertical string insulators. *Figure 16* shows a front- and back-guyed single circuit wood H-frame dead-end structure. The single circuit phase conductors are supported by tension string insulators. *Figure 17* shows a modified wood H-frame suspension structure using three string insulators and two rigid insulators for support of the three-phase conductors along with a center-span spacer using rigid insulators.

3.0.0 General Transmission Line Construction Requirements

Power authorities and/or utilities must go through a long approval process to construct new transmission lines. Many federal, state, regional, and local agencies/jurisdictions may be involved in the approval process. Part of the process requires that the entire transmission line path must be inspected to determine accessibility, work-site hazards, environmental concerns, and right-of-way access through public and private property.

Once approval is obtained, many power authorities and/or utilities subcontract the actual construction for major projects. This involves contracts with detailed specifications and drawings for the construction of the transmission lines. In-house or subcontracted engineering is used to plan and detail the contract specifications and drawings as well as to survey and stake the transmission path and monitor the work being accom-

81202-11_F12.EPS

Figure 12 345kV self-supporting coated single-circuit dead-end TSPs.

81202-11_F13.EPS

Figure 13 Self-supporting galvanized/coated TSP single-circuit H-type suspension structure.

plished. Most of the detailed specification and construction requirements provided in a contract cover the following:

- Access
- Right-of-way clearing
- Structure foundations
- Structure framing/erection
- Guys and anchors, as required
- Electrical grounding, bonding, and testing

Specifications and construction requirements including example drawings for construction of wood H-frame tangent structures are discussed later in this module. The example wood structure drawings are from the US Department of Agriculture (USDA) Rural Utilities Services (RUS) *Bulletin 1728F-811, Electric Transmission Specifications and Drawings, 115kV Through 230kV.*

81202-11_F14.EPS

Figure 14 Self-supporting wood H-frame single-circuit suspension structure.

In addition to detailed transmission path specifications and construction requirements, contracts also require contractors to observe the following:

- Adhere to specified work schedules.
- Adhere to transmission path maps and verify the location of structures.
- Avoid damage to property by all personnel engaged in the work.
- Follow all federal, state, and local electrical codes if more stringent than the contract specifications.
- Follow all federal, state, and local safety laws, regulations, and codes, and observe proper safety practices. This may include developing, maintaining, and monitoring a safety program.

3.1.0 Access

Access to a transmission line path usually involves the creation of non-public roads. Transmission line roads are classified as either access roads or spur roads. Access roads are through-roads that run between tower sites alongside of or inside the transmission path right-of-way (ROW), as shown in *Figure 18A*. Spur roads, if needed, are roads that lead from access roads and terminate at one or more transmission structure sites. Part of a short spur road is shown at the bottom of *Figure 18A*. *Figure 18B* shows an access road in a ROW that crosses and is accessible from a public road. They are generally blocked with a locked gate. *Figure 18C* shows a temporary access road crossing private property. *Figure 18D* shows the construction of a rough terrain access road.

Points of access to the ROW across private property for the purposes of construction and, in some cases, for future maintenance of the transmission line, require documented, prior approval from the property owners by either the contract engineers or the contractor as specified in the contract. The documented approval should include releases for any damage to the property. Access across public land must be obtained from the authority having jurisdiction (AHJ).

If temporary and/or permanent access roads (*Figure 18*) are required as part of the work, roads must be of the type, dimensions, and grades as specified in the contract. They must be located as shown on the contract plans and as staked by the engineers. The contractor must use compactable granular material for fill as specified in the contract and the fill must be compacted for use by heavy equipment. As specified in the contract, culverts, water bars, and drainage ditches must be installed in accordance with the plans and specifications. For temporary access roads, the contractor is required to restore land that is crossed for this purpose to its original condition.

If fences must be cut to allow access for the work, gates must be installed as specified in the contract drawings or by the engineers. Fence gates must be closed and locked when required by the landowner. Unless certain access points remain for line maintenance, gates are normally removed and the fence restored at the completion of the project.

3.2.0 Right-of-Way Clearing

Clearing of the ROW for a transmission line may involve the full width clearing, selective clearing, tree topping, or other forms of ROW preparation. *Figure 19* shows a cleared transmission line path prior to erection of the line. Only those areas shown on path plans and/or drawings or specified by engineers should be cleared. *Figure 20* shows a typical ROW clearing guide used for wood H-frame structures. The ROW width is based on the structure height and width dictated by the voltage

81202-11_F15.EPS

Figure 15 Side-guyed, three-pole single-circuit wood deviation/tension structure.

81202-11_F16.EPS

Figure 16 Front- and back-guyed wood H-frame dead-end structure.

81202-11_F17.EPS

Figure 17 Modified wood H-frame suspension structure.

being transmitted. Normally, trees outside of the ROW that present a danger to transmission lines are marked for removal by engineers. Vegetation should only be removed as necessary to allow construction, operation, and maintenance of the transmission line. Care must be taken to prevent removal of ground cover, which could lead to soil erosion. Recent federal laws and regulations for federal lands require selective clearing. This means that low-growing trees and brush that will not interfere with the transmission line in the future should not be cleared. States and other authorities may also require selective clearing.

Normally, trees that must be removed in the ROW are felled and the stumps are removed. However, if stumps are allowed by the contract to be left in the ROW for certain areas because of inaccessibility by wheeled or tracked equipment, they have to be treated with a heavy application of an appropriate herbicide to prevent regrowth. Herbicides can only be used with the approval of an engineer and only in areas designated for their use. The herbicides must meet the environmental requirements of all governing agencies. They must be applied in accordance with the manufacturer's recommendations and only by a licensed/certified applicator.

4.0.0 Structure Foundations

With the exception of directly buried wood transmission structures and some concrete/synthetic material tubular poles, LSTs and many TSPs are erected using various types of foundations for support. Soil and subsurface composition is the major factor in determining the type of foundations that are selected by engineers and specified for a project along a transmission line path. Before any foundations are constructed at a structure site, the site must be graded and cleared to provide a reasonably level and vegetation-free surface. The grading should be such that surface water runs toward the direction of natural drainage to eliminate ponding and prevent erosive water flows that could damage foundations. The graded area should be compacted to at least 90 percent relative density and should be able to support heavy vehicular traffic. If the site must accommodate structure framing in addition to structure erection, a considerable amount of land must be prepared at each structure site depending on the size of the structure. In mountainous areas, benching may be required for structure construction operations as well as for some structure foundations. Benching is a technique for creating terraced land in extremely steep and rugged terrain.

Some of the foundations that have been or are being used for LSTs are shown in *Figure 21*. In most cases, a fabricated template made of light angle iron or square tubing can be used to represent the angle and spacing of the legs at the bottom of the structure to provide the correct spacing and angle of anchors. The anchors are bolted to the template and are held in the proper position until concrete

Figure 18 Access roads.

81202-11_F19.EPS

Figure 19 A cleared transmission line path.

is placed and has set; or, in the case of directly buried anchors, until the soil is replaced and properly compacted. *Figures 21A* and *B* show tapered concrete foundations with an embedded anchor-bolt cage or an angle-iron stub with a bottom plate. *Figure 21C* shows a rock anchor embedded in concrete. *Figures 21D, E,* and *F* show galvanized angle-iron stub anchors buried directly in the soil. Various methods of forming underground anchor configurations for the stubs, including grillage, are shown with the tower leg connected to the stub either above or below ground. Directly buried angle-iron stub anchors can also function as earth grounds for all four legs of a structure. *Figure 22* shows a directly buried angle-iron stub anchor for an LST leg. The leg and bottom diagonals are bolted to the heavily galvanized ⅜-inch thick angle-iron stub anchor. *Figure 23* shows an LST leg connected to a tapered concrete foundation with an embedded angle-iron stub. Note the two radial cable grounds connected to the LST leg.

Large LSTs and TSPs generally use concrete foundations. *Figure 24* shows a TSP and LST supported by a concrete foundation with an anchor bolt cage embedded in the foundation. The flange on the TSP is fastened to the anchor bolts by nuts when the structure is erected. The LST foundation leg has a three-bolt anchor cage embedded in the concrete. The mounting pads welded to the bottom of the leg are fastened to the three-bolt anchor cage with double nuts. Note the ground wire at the base of the leg going to an external buried ground rod. For typical soil compositions, these types of foundations are constructed as drilled pier foundations. *Figure 25* shows the anchor bolt cage aligned and leveled in a drilled hole and in position for concrete placement. A long handheld air-powered vibrating wand is used during concrete placement to eliminate voids and air bubbles. After placement, the surface of the foundation is smoothed and finished. The template shown at the top of the cage is removed after the concrete has cured.

NOTE

The concrete form in this application is removable. In other cases, permanent metal or cardboard forms are used and are left in the ground.

Drilled pier foundations for each LST leg are normally 3 to 6 feet in diameter and 15 to 30 feet deep. For TSPs, the drilled pier foundations can be up 10 feet in diameter and up to 60 feet deep. One disadvantage of concrete foundations is that it takes a minimum of 20 days for the concrete to cure before a structure can be erected. This means that foundation crews must work well ahead of any structure erection crews.

Other types of foundations are used for differing types of soil or subsurface conditions and/or for inaccessible terrain. *Figure 26* shows part of one type of process for construction of a concrete LST leg foundation in terrain with a rock subsurface. *Figure 27* shows a portable drilling rig boring holes in a rock subsurface

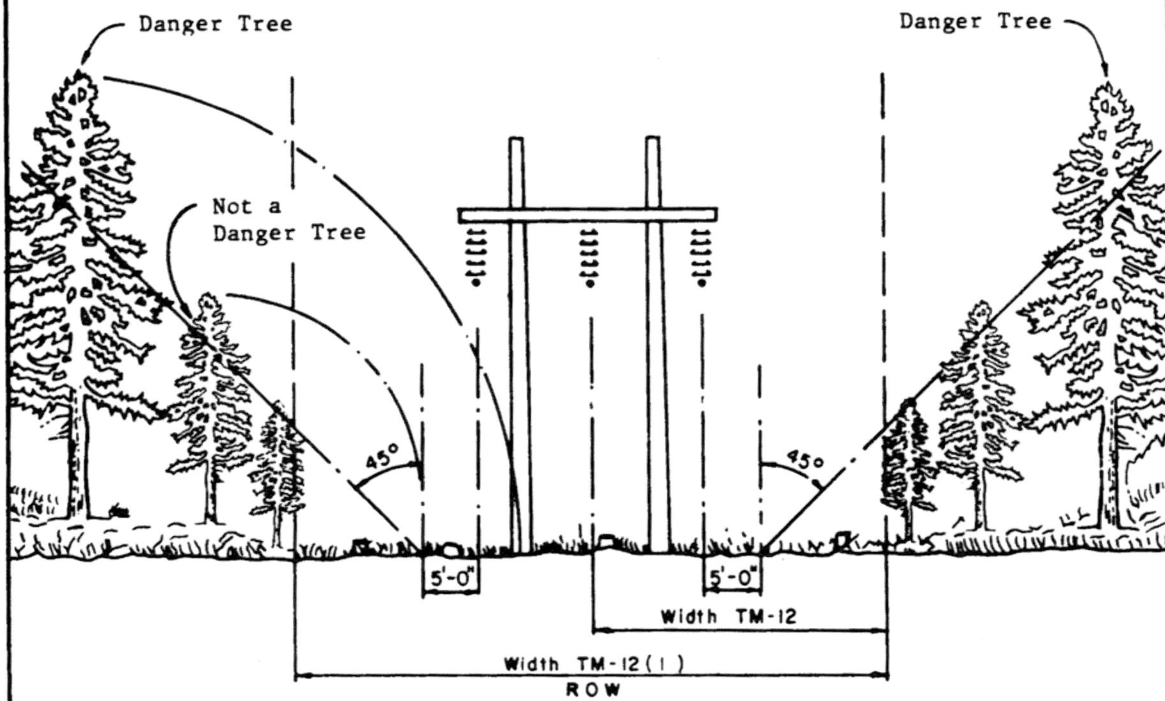

NOTES:

1. Engineer will designate all danger trees which shall be removed or topped at option of contractor. In approximately level terrain, trees which would reach within 5 feet of a point underneath the outside conductor in falling are examples of danger trees.

2. As directed by the engineer, portions of the right-of-way (ROW) must be cut so that stumps will not prevent the passage of tractor and trucks along the ROW.

3. The unit for clearing one-half of the ROW is "WIDTH TM-12."

4. The unit for clearing the full ROW is "WIDTH TM-12(1)."

5. The unit for clearing danger trees is "TM-13."

TRANSMISSION ROW CLEARING

RIGHT-OF-WAY CLEARING GUIDE

NO.	REVISION	DATE	Aug., 1986		TM-12,-12(1),-13

Reissued 03/98

81202-11_F20.EPS

Figure 20 Typical ROW clearing guide for a wood H-frame structure.

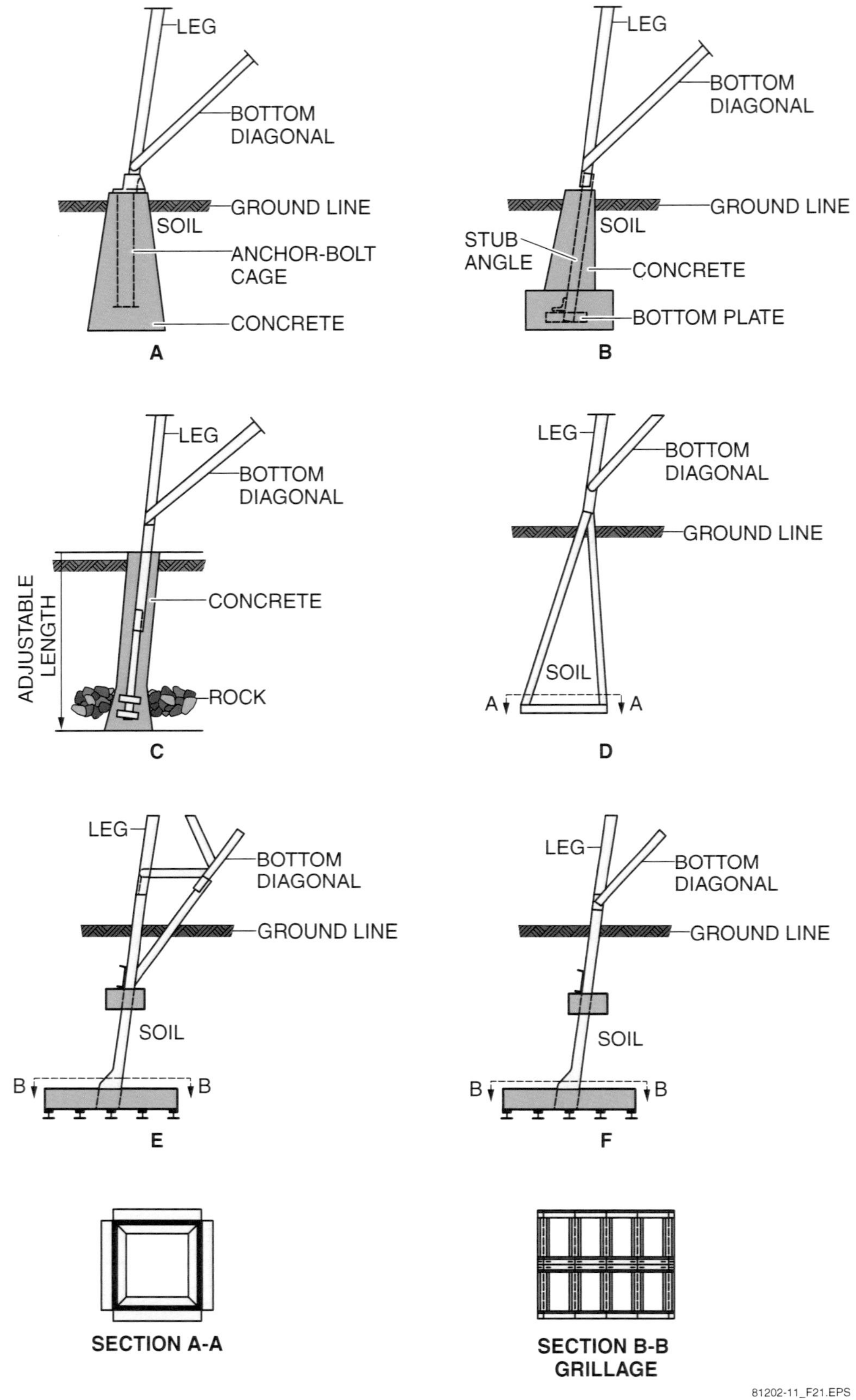

Figure 21 LST foundations.

81202-11_F22.EPS

Figure 22 Directly buried angle-iron stub anchor for an LST leg.

81202-11_F23.EPS

Figure 23 LST leg tapered concrete foundation.

81202-11_F24.EPS

Figure 24 TSP and LST supported by anchor bolts in concrete foundations.

for an LST leg. The holes are for the installation of grouted micropile foundation anchors in inaccessible terrain. The anchors will be capped with a reinforced concrete pad containing anchor bolts for the structure leg. For shorter foundation crew lead times, various non-concrete power-driven foundation methods can be used. *Figure 28* shows a tubular steel pile being driven to support one leg of a tubular steel X-frame transmission structure. In inaccessible areas, portable vibrating pile drivers transported by helicopter are clamped to the pile and used to drive it into place. *Figure 29* shows a group of four screw pilings driven at angles for support of an LST leg in unstable soil. The tops have been cut off and a mounting plate has been welded to the top of them to support the structure leg. An angle-iron stub has been welded to the plate for bolt-attachment of the structure leg and bottom diagonal. These types of pilings are driven by a rotating high-torque hydraulic driver attached to an articulated hydraulic crane boom. The pilings are available in hollow steel pipe increments from 2 to 16 inches in diameter and up to 36-feet long.

81202-11_F25.EPS

Figure 25 TSP anchor bolt cage prepared for concreting.

5.0.0 Structure Framing and Erection

Because most structures are custom fabricated for size and strength as specified by the project engineers, the framing or assembly drawings vary considerably due to the way the fabricators design the structures to meet the specifications. Besides structure framing and erection, a transmission line project requires other important elements. Depending on the extent and terrain for a transmission path, one or more primary marshalling yards may be established. Materials and equipment typically staged at these yards would include, but not be limited to, unassembled structure components, wire reels, insulators and hardware, heavy equipment, light trucks, construction trailers, and portable sanitation facilities. An area of five acres or more would typically be required for a primary yard. In addition to construction materials and equipment, these yards may contain trash and recycle bins. The marshalling yards are normally surrounded by perimeter fencing as well. In addition to primary yards, secondary marshalling yards may also be required for short-term use nearer to the construction sites. They typically require an area of one to three acres.

81202-11_F26.EPS

Figure 26 LST leg foundation construction in a rock subsurface.

81202-11_F27.EPS

Figure 27 Micropile construction for an LST leg in an inaccessible location.

81202-11_F28.EPS

Figure 28 Driving a tubular steel pile for an X-frame transmission structure.

81202-11_F29.EPS

Figure 29 Screw-type pilings.

5.1.0 Examples of Structure Framing and Erection

Framing of LSTs and TSPs can be accomplished in marshalling yards or at the construction sites depending on the size of the structures and the location's terrain. For LSTs, the framing for sections of the structure can be done in a yard. The sections are then transported to a structure construction site by truck or helicopter. In other cases, framing may be done completely at the structure construction site from bundles of unassembled components. In the case of TSPs, they may be completely framed in a yard and transported to the structure site by helicopters or transported to the construction site as unassembled pole sections with final framing prior to or as part of the erection process. For both LSTs and TSPs, the framing/torquing crew or the erection/torquing crew, as applicable, must make sure that each bolt, nut, washer, or lockwasher for a structure is properly installed. The crew must also ensure that the nuts and bolts are torqued to the correct value or premature failure of the structure can occur. Framing or erection crews may also elect to install insulators and transmission line rollers (travelers). If this is to be done, it can be accomplished either as part of the framing operation or erection operation.

Figure 30 shows the framing, transport, and erection of a large V-type 765kV single-circuit suspension LST. Framing of the tower legs and a large portion of the bridges for this particular project is accomplished at a marshalling yard and the assembled sections are stored at the yard as shown in *Figure 30A*. In *Figure 30B*, a helicopter crane has delivered the V-type tower legs (with a temporary bridge) to the construction site. It remains and hovers until the base of the structure legs is secured to its foundation and the structure's four supporting guy wires are temporarily secured to four guy-wire anchors. The guy-wire anchors are located at four corners a considerable distance from the tower base. After the guy wires have been adjusted and tensioned enough so that the V-leg structure is vertical and stable, the bridge structure is installed. *Figure 30C* shows the delivery of the bridge structure that is bolted to the top of the V-leg structure in place of the temporary bridge. Eventually, the remaining parts of the bridge are airlifted and installed on the bridge. Then, after stringing operations, the guy wires are readjusted for final position of the structure and tightened to tension specifications.

81202-11_F30.EPS

Figure 30 Framing, transport, and erection of a large V-type 765kV single-circuit suspension LST.

Figure 31 shows the erection of a non-sectioned weathering-steel single-circuit suspension TSP. It is replacing an old LST that has had two of its legs cut and lowered. In *Figure 31A* a hole is being drilled for direct burial of the pole butt in the center area of the old LST. In *Figure 31B* the davit arms and string insulators have been installed and a cable for lifting is being connected to a hydraulic-boom crane hook. *Figure 31C* shows the beginning of the lift and *Figure 31D* shows the lift nearly completed. In *Figure 31E,* the TSP has been lifted off the ground in preparation for insertion in the drilled hole. *Figure 31F* shows the pole inserted in the hole and soil being replaced around the pole. The two crew members to the right of the pole are preparing an air-powered tamping tool to compact the soil. Note that the bottom section of the pole is covered with a factory-applied protective coating to prolong the life of the pole.

Figure 32 is an example drawing of a framing layout area at a construction site for a large, heavy single-circuit LST. The layout area of 200' by 200' is based on a structure height of 185 feet. Bundles of structure components are brought in by truck and are stacked at their respective positions for framing. A proper capacity-rated truck crane is placed at a near-center position in order to reach all the assembled sections and the structure location for erection operations. Note the weight of the assembled sections for this tower.

81202-11_F31.EPS

Figure 31 Erection of a non-sectioned weathering-steel single-circuit suspension TSP.

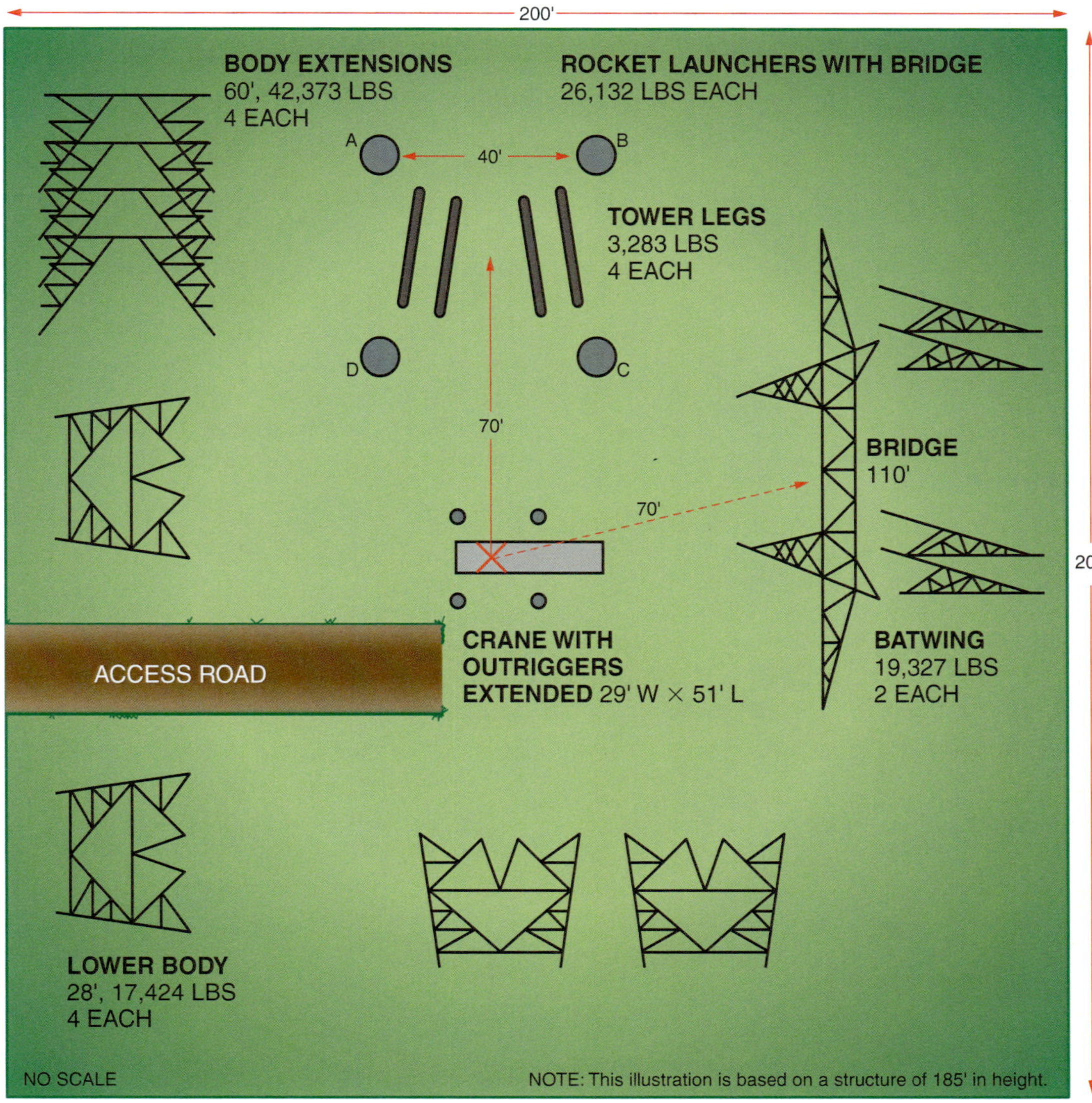

Figure 32 Example framing layout area drawing of a structure construction site.

A number of TSP and LST structures are shown in *Figures 33* through *35* in various states of erection. *Figure 33* shows a large single-circuit TSP supported by a crane before being bolted down to its foundation. *Figure 34* shows a hot-line erection of a replacement weathering-steel H-type single-circuit TSP structure. Great care must be taken when work of this type is performed in close proximity to energized high-voltage lines. Spotters must be used at all times for bucket truck or crane operations. In *Figure 34A* the base section of an H-pole is being lowered into a drilled hole for direct burial. Note the factory applied protective coating on the portion of the pole that will be buried. The upper section of the pole is in the foreground. After the other base section is erected, the upper sections that support the cross-arm are installed. *Figures 34B* and *34C* show both upper pole sections in place and the cross-arm being installed. Note that the hot lines are positioned such that the cross-arm can be carefully maneuvered into position and secured. *Figure 34D* shows the completed H-type TSP structure with the top section for the earth-ground wires installed, the hot wires transferred, and the old LST removed. *Figure 35* shows the erection of a two-circuit 500kV trellis-type LST. *Figure 35A* shows a section being moved into position so that the four crew members atop the previously erected sections can secure it in place. *Figure 35B* shows the last support arm being installed and *Figure 35C* shows the completed structure before stringing operations. Note the benched construction area as well as the benched concrete tower leg foundations at each side of the tower.

81202-11_F33.EPS

Figure 33 Large single-circuit TSP to be bolted to its foundation.

5.2.0 Framing and Erection of a Wood H-Frame Transmission Structure

Detail information for the framing and erection of a wood H-frame suspension (tangent) structure is covered in *Appendix A* and *Appendix B*. *Appendix A* provides an example contractor specification for the portion of the contractual work to erect wood H-frame structures. *Appendix B* provides the framing and detail drawings, including a cross-arm drilling drawing, for a 230kV tangent wood H-frame transmission (TM) structure. Pole sizes and lengths as well as string insulators and their hardware are specified in the contract by an engineer. In most cases, poles and cross arms are predrilled for erection by a supplier. For tall poles, a double X-brace may be required and is specified in the contract as a TM-230X structure as identified on the TM-230 drawing. Framing of an H-frame structure is normally done in a layout area at the structure site location. When the framing is completed, it should be inspected to make certain all unused drilled holes are treated with preservative and plugged properly; all components are installed correctly and treated with preservative; and all bolts/nuts are tightened properly. The bottom holes for the X-brace(s) are not drilled until the assembled structure is erected and placed in post holes using a crane. The steel hardware used for framing, the wood drilling required, and post-hole boring is the same or very similar to that of wood distribution poles.

Figure 34 Hot-wire erection of a replacement H-type single-circuit TSP structure.

Figure 35 Erection of a two-circuit 500kV trellis-type LST.

Figure 36 shows a structure very similar in construction to the one shown in the appendices. *Figure 37* covers some of the details of that structure. Note that the cross-arms appear to show signs of deterioration. *Figure 37A* shows a view of the double cross-arm and its attachment to a pole. Flat washers appear to have been used between the cross-arms and the pole. *Figure 37B* is a close-up view of the outer brace for the cross-arm and the bracket for the brace and insulator. *Figure 37C* shows the detail of the cross-arm braces, attachments, and earth-grounding arrangement at the top of one of the poles. *Figure 37D* shows the cross-arm center braces and the bracket for the braces and center insulator. Note that the difference between this structure and the TM-230 drawing is that the steel tie-angle at the top and between the two poles on the drawing is replaced by a tensioned pole-tie cable in this structure.

Figure 36 Wood H-frame tangent structure.

6.0.0 Guys and Anchors

For transmission lines, only structures that are not self-supporting require guys and anchors. Structures such as Y-type, X-type, or V-type towers and shallow- or large-angle wood structures need anchors and guy lines. A deviation/tension TSP may need anchors and guy lines if it not self-supporting. In any case, installation of

81202-11_F37.EPS

Figure 37 Details of a wood H-frame tangent structure.

anchors and guy lines parallel the information given in earlier modules. Guys and anchors must be installed at the locations shown on plans and drawings or as specified by the project engineers. Anchor rod locations must be staked by an engineer and the contractor should check the locations before structure construction. Some general installation requirements are as follows:

- Anchor rods must be installed in line with the guy wire and must be installed so that not more than 8 inches of rod, including the eye, remain out of the ground after tension is applied. In cultivated land, exposure may be increased to 12 inches to prevent burial of the rod eye.
- Anchors must be the size, type, and length as specified in the plans or drawings.
- Anchors in excavated anchor holes must be approved by an engineer before concreting or backfilling and tamping the hole. Only approved granular material or native soil can be used for backfill.
- When required by the contract, anchors must be tested to 50 percent of their designated ultimate rated capacity.
- Power-driven screw anchors must be installed with the appropriate equipment. Screw anchors must not be reversed to meet the rod projections above ground.
- Guys must be pre-tensioned to eliminate any slack before the transmission lines and overhead earth ground wires are strung. Guys may have to be retensioned after stringing operations are completed to plumb the structures and equalize guy tensions. The final tension of the guys must meet the approval of a project engineer.

Some of the anchors used are shown in *Figures 38* and *39*. An installation method for an expanding anchor (bust-type) and equipment for driving a screw anchor are shown in *Figure 40*. Today, the most common anchors are screw anchors. Most of the screw anchors are power driven and are available with ratings up to 200,000 pounds minimum ultimate tension strength with a maximum installation torque rating of 20,000 ft-lbs. *Figure 41* shows an adjustable termination grip used for the bottom of a guy wire. A non-adjustable version is used for the top of the guy wire. They are available for guy wire sizes up to one inch in diameter for galvanized wires with a load rating up to 104,500 pounds. They are also available for aluminum-coated guy strands up to 1.270 inches in diameter with a load rating up to 142,900 pounds.

Figure 38 An expansion anchor and tool.

7.0.0 Electrical Grounding, Bonding, and Testing

Earth grounding of transmission structures and overhead ground wires is necessary to reduce damage by lightning strikes. Overhead ground wires are used to intercept lightning before it strikes the transmission lines between the transmission structures. If the impulse ground resistance of the transmission structures and overhead ground wires is high, flashover to the transmission lines can occur,

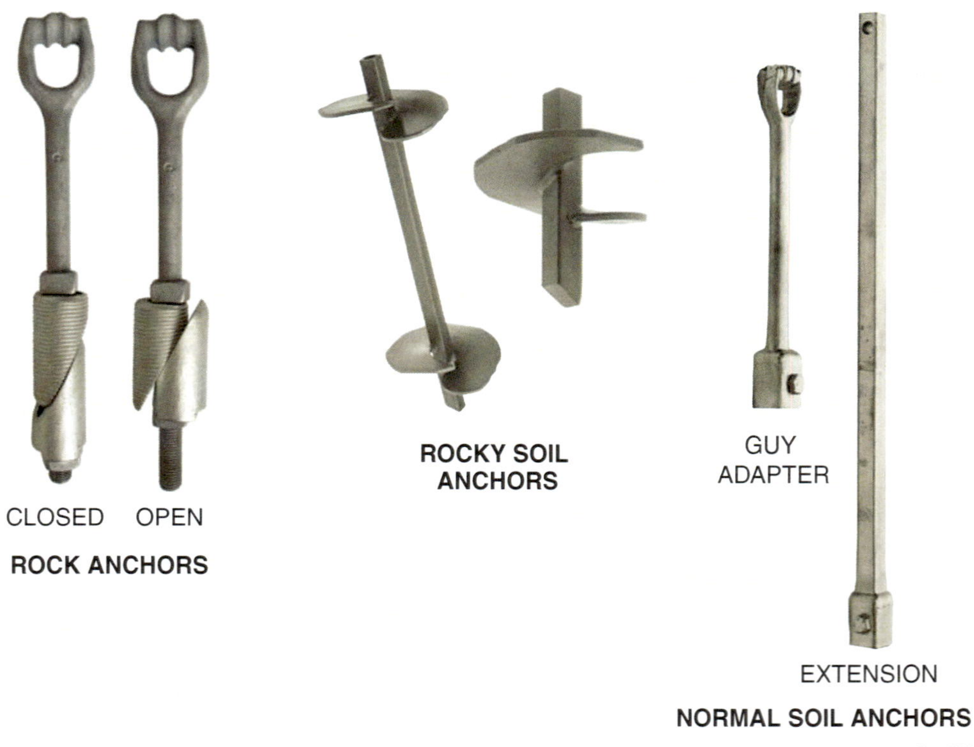

Figure 39 Various anchors.

knocking the lines out of service. Transmission structure grounds must be installed as specified in the contract. Ground wire should be solid or stranded copper wire only. Braided copper wire should not be used. Normally, earth ground rods must be driven 8' to 10' deep for a proper ground. Threaded sectioned steel ground rods with a thick layer of copper bonded to the steel are available for this purpose. The steel core along with a steel driving point and a steel driving head with a replaceable driving stud allow the rod to be driven to well beyond the normal 8- to 10-foot depth of a solid-

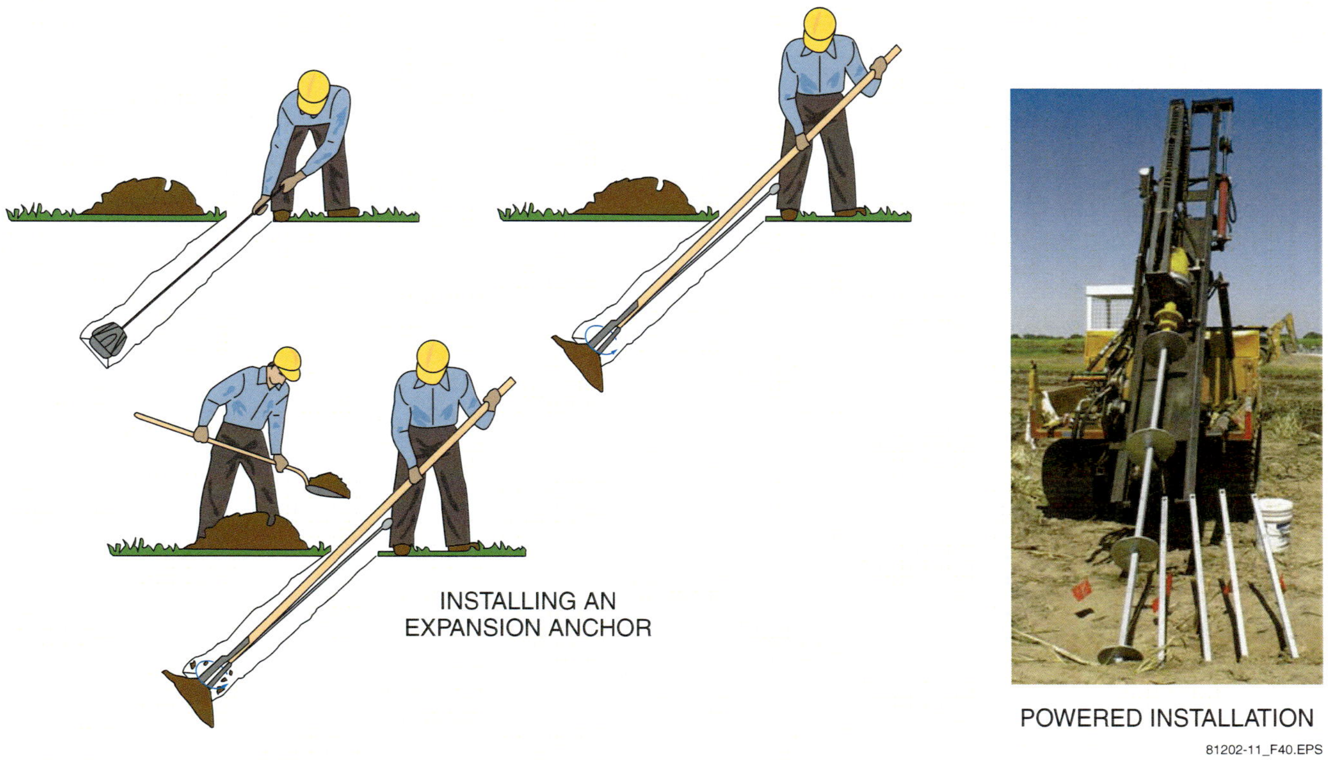

Figure 40 Installation of an expansion anchor and power equipment for screw anchor installation.

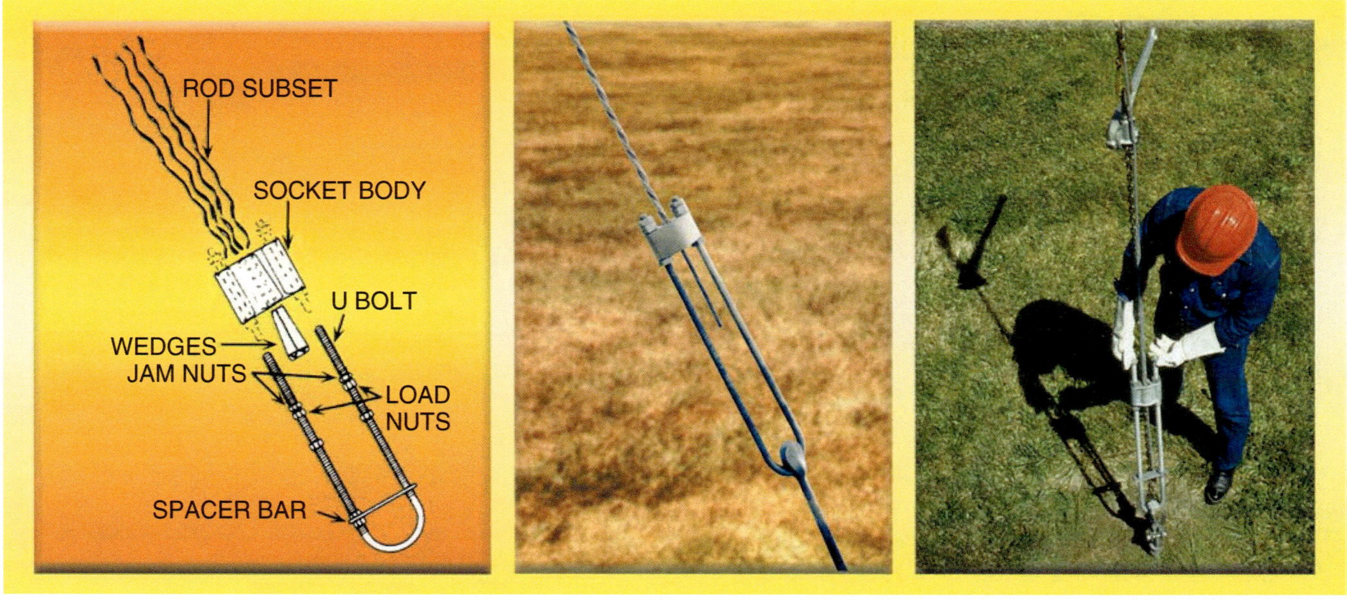

Figure 41 Termination grip for guy wires.

copper ground rod. The sections are fastened together as required by aluminum bronze couplings and the steel driving head is transferred from the top of the last section to the top of the new section to continue the driving process. Some contracts specify that instead of mechanical methods of connecting ground wires together or to structures or ground rods, earth ground wiring must be connected by exothermic welding (Cadweld) and/or by implosive connectors.

For wooden poles, the last two pages of *Appendix B* show the methods used to ground the overhead earth ground wires. In addition to those methods, a spiral winding of solid copper ground wire can be stapled to the butt of the pole or a copper butt plate (*Figure 42*) may be used to help insure grounding.

Earth grounding of protectively coated TSPs is shown in *Figure 43* using single ground rods. The ground rods shown in the figure are normally buried.

Earth grounding of LSTs must be accomplished at all legs. If galvanized steel stub anchors are directly buried and are 8 to 10 feet below ground level, they are essentially the earth ground in most soils. If they are bolted to angle-iron stubs embedded in the concrete foundations, then they have to be grounded using external ground rods unless grounds have been created for the angle-iron stubs before the concrete was placed.

81202-11_F42.EPS

Figure 42 Copper butt plate.

Resistance testing of grounds must be done to make sure the ground resistance is below 25 ohms or as specified in the contract. A variety of ground resistance-measuring devices are available. Standard ground resistance measurements must be made before the stringing of overhead ground wires or the addition of counterpoise. Counterpoise is a term applied to the creation of a grid for additional grounding capability in high-resistance soil areas. *Figure 44* shows an AEMC Instruments Model 6472 and Model 6474. The 6472 unit makes several different conventional ground measurements but, coupled with the 6474 unit, it can make simultaneous measurements of four-legged or two-legged LSTs or one- or two-pole TSPs, even though they may be connected to buried counterpoise and/or overhead ground wires.

81202-11_F43.EPS

Figure 43 TSP earth grounding.

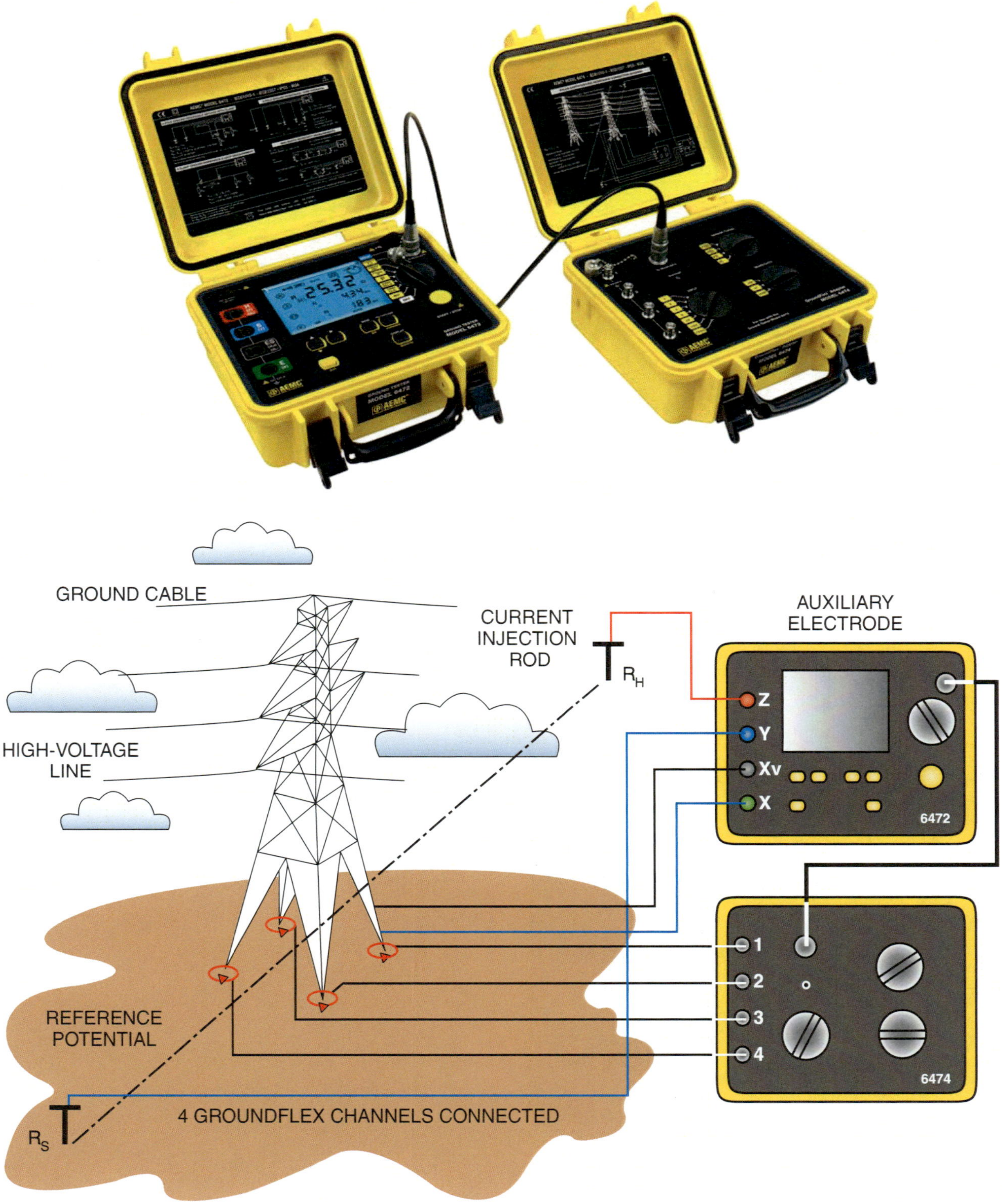

Figure 44 Ground tester and adapter for tower grounds.

If ground resistance is not satisfactory, counterpoise may be added if approved by the project engineers. Counterpoise may be accomplished by various methods including those shown in *Figure 45*. *Figure 46* shows an LST leg with one of the radial grounds damaged and uprooted. This could result in unbalanced grounding of the leg relative to the other legs.

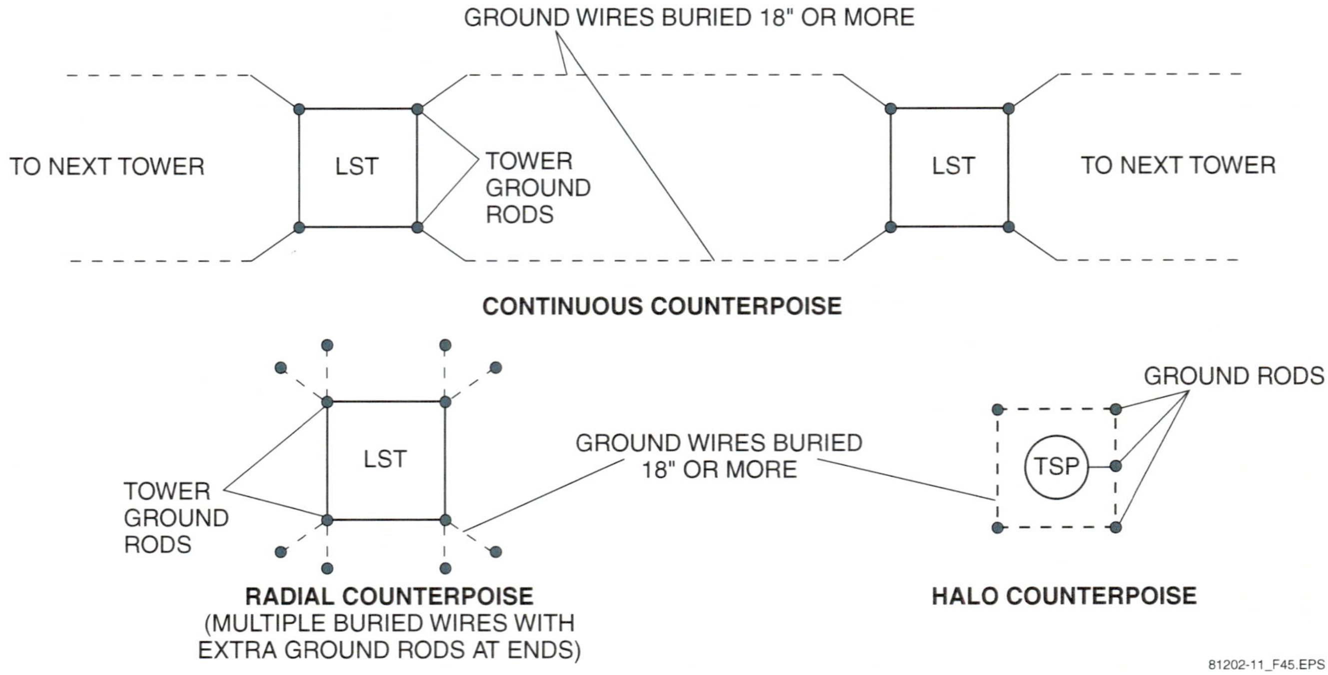

Figure 45 Counterpoise methods.

Figure 46 An uprooted and damaged LST leg radial ground.

SUMMARY

The purpose of transmission structures is to supply high-voltage electricity within and between regional power networks using overhead transmission lines. These transmission lines normally operate at voltages ranging from 69.5kV to 765kV. Transmission structures vary greatly in types, designs, and sizes depending on the terrain, soil conditions, and the voltage/current being transmitted. They can consist of wood, lattice steel, or tubular structures. Lattice steel or tubular structures are commonly called towers.

Construction of transmission structures is usually accomplished by contractors. Framing (assembly) and erection of the towers vary depending on terrain along the construction path of the transmission line and must be accomplished in accordance with contract specifications and drawings. Some construction sites along the transmission path may be accessible by wheeled or tracked vehicles and others may be accessible only by helicopter cranes. For accessible sites, framing and erection can be accomplished at the structure construction sites. Framing for accessible sites can also be accomplished at marshalling yards with transport to the construction site by truck or helicopter. For inaccessible sites, framing can usually only be accomplished at marshalling yards with transport and erection of the towers at the construction site accomplished with the aid of helicopters.

The construction path of the lines may require right-of-way access. This entails the proper construction of any required access roads in accordance with contract specifications and drawings. Proper clearance of the path to remove certain vegetation must be accomplished in accordance with contract specifications. Proper foundations, guy wire anchors, and electrical grounding, bonding, and testing for the transmission structures must also be accomplished in accordance with contract specifications and drawings.

Review Questions

1. An ironman is another term for a(n) _____.
   a. LST structure
   b. suspension structure
   c. load structure
   d. deviation/tension structure

2. Terminal structures are also called _____.
   a. dead-end structures
   b. tangent structures
   c. load structures
   d. angle structures

3. In *Figure 1*, D and F are called _____.
   a. load string insulators
   b. tension string insulators
   c. anchor string insulators
   d. vertical string insulators

4. In *Figure 1*, the components labeled E are called _____.
   a. load string insulators
   b. tension string insulators
   c. anchor string insulators
   d. vertical string insulators

5. *Figure 2* shows a(n) _____.
   a. bridge-type suspension tower
   b. V-type suspension tower
   c. trellis-type suspension tower
   d. H-type suspension tower

81202-11_RQ02.EPS

Figure 2

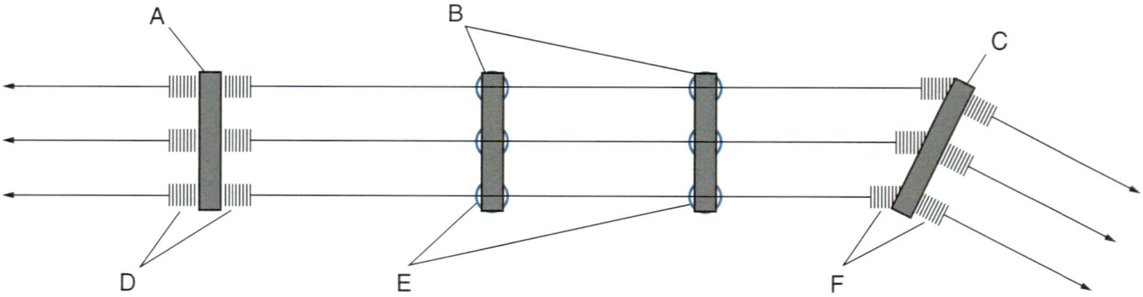

81202-11_RQ01.EPS

Figure 1

81202-11_RQ03.EPS

Figure 3

6. *Figure 3* shows LST _____.
   a. bridge-type angle structures
   b. V-type angle structures
   c. trellis-type angle structures
   d. H-type angle structures

81202-11_RQ04.EPS

Figure 4

7. *Figure 4* shows a(n) _____.
   a. bridge-type suspension structure
   b. V-type suspension structure
   c. trellis-type suspension structure
   d. H-type suspension structure

81202-11_RQ05.EPS

Figure 5

8. *Figure 5* shows a wood _____.
   a. bridge-type suspension structure
   b. V-type suspension structure
   c. trellis-type suspension structure
   d. single-circuit suspension structure

9. The requirement for a safety program is normally specified for a project in the _____.
   a. specifications
   b. contract
   c. drawings
   d. execution documents

10. A road that leads from an access road to one or more transmission structure sites is called a _____.
    a. side road
    b. dead end road
    c. public road
    d. spur road

11. Selective clearing is currently required for all _____.
    a. private land
    b. state land
    c. federal land
    d. local public land

12. Graded structure construction sites should be compacted to a relative density of _____.
    a. 60 percent
    b. 70 percent
    c. 80 percent
    d. 90 percent

13. LST drilled pier foundations for each leg are typically _____.
    a. 1 to 4 feet in diameter
    b. 3 to 6 feet in diameter
    c. 5 to 8 feet in diameter
    d. 7 to 10 feet in diameter

14. V-type TSPs are anchored at _____.
    a. two points
    b. three points
    c. four points
    d. five points

15. Grounding of transmission structures is necessary to _____.
    a. reduce damage by lightning strikes
    b. increase induced voltage
    c. reduce electrolytic corrosion of the structure
    d. provide a path for flashover to ground

Trade Terms Introduced in This Module

Circuit: Any set of three-phase transmission lines used on a structure.

Dead-end structure: Normally used at the endpoints of specified lengths of line conductors after a specified number of spans between suspension structures.

Deviation/tension structure: Used to sharply change the direction of transmission lines horizontally or on inclines.

Lattice steel tower (LST): A transmission structure made up of numerous steel strips and angles.

Suspension structure: Used along relatively straight and level runs of transmission lines. They are also used for long shallow turns.

Transmission structure: Used to support overhead lines (conductors) for electric power transmission.

Tubular steel pole (TSP): A transmission structure that uses round steel members.

Appendix A

Portion of Specifications for Wood H-Frame Structures

4. WOOD POLES

4.1 Pole Inspection, Handling, and Distribution

4.1.1 The contractor shall immediately notify the engineer of freight damage or misfabrication of poles. The framing, boring, and gaining, if required, must agree with the Pole Framing Details for the specific structures to be erected.

4.1.2 The pole lengths and classes must agree with the Pole Units specified for the structures to be erected, as tabulated in the Transmission Construction Units and shown on the plan-and-profile drawings.

4.1.3 Poles must be handled with care so as not to damage the wood or preservative treatment. Poles must be lifted off of the pole hauler at designated structure locations. They must not be rolled or dragged along the ground. Lifting slings must be used and must be fabricated from a material that protects the wood from damage. Pole tongs are not permitted when handling poles.

4.1.4 If poles are stored after delivery, they must be carefully arranged and placed on wide blocking to prevent crushing. Poles must not come in contact with standing water or the ground. No pole will have an unsupported length greater than 20 feet. The blocking must be provided by the contractor and included in this unit price.

4.1.5 The contractor shall distribute extra heavy, choice, close-grained poles to angle, deadend, and crossing structures.

4.1.6 When ungained poles and adjustable spacer fittings are specified for multi-pole structures, the contractor shall match the poles as directed by the engineer.

4.2 Field Drilling and Treating

4.2.1 All field drilled holes must be thoroughly treated with a heavy application of preservative compound approved by the engineer. The contractor shall include any required field drilling and treating in the unit cost for pole top assemblies.

4.2.2 Unused holes or holes that are misdrilled must be plugged prior to erection using treated wood dowel pins 3 inches in length. When a hole is misdrilled, the engineer shall be notified. A pole will be rejected by the engineer if two or more misdrilled holes occur at a connection. If a pole is rejected due to misdrilling of holes by the contractor, the contractor shall replace the damaged pole at no additional cost to the borrower.

4.2.3 Double crossarm spacer fittings, if required, are indicated in the List of Materials on the transmission line

81202-11_A01.EPS

structure drawings. Fixed spacers must be used with factory gained and treated poles. Adjustable spacers must be used with ungained and treated poles. The gaining of poles, and the type of crossarm spacers are as follows (engineer to check one):

a. Factory gained and treated poles for fixed spacer fittings ....☐

b. Ungained and treated poles for adjustable spacer fittings ....☐

c. Gained and treated poles for adjustable spacer fittings ....☐

d. Gained and treated poles for adjustable spacer fittings ....☐

d. Structure with double crossarms are not required ..☐

4.2.4 Poles may be supplied with flat or 15 degree sloping roofs. Tops of poles must not be cut except with the expressed approval of the borrower. If cutting is deemed necessary, the pole top must be cut off at a slope of 15 degrees, thoroughly treated with preservative approved by the engineer, and covered with a mastic type cap. Poles are to be supplied with (engineer to check one):

a. Flat roofs ....☐

b. 15 degree sloping roofs ....☐

c. Contractor's option ....☐

4.2.5 The butt of any pole is not to be cut under any circumstances.

4.3 <u>**Special Requirements**</u> (to be completed by the engineer)

81202-11_A02.EPS

5. **POLE TOP ASSEMBLIES**

5.1 **Reference to Drawings**

5.1.1 The pole top assembly unit consists of all items shown in the List of Materials on the transmission line structure drawings. If spacers for double crossarms are required, the type of spacer must be as noted in Wood Poles, section 4.2.3.

5.1.2 Unless shown in the List of Materials, the pole top assembly unit does not include other units such as pole units, pole grounding units, foundation units, guying assembly units, and anchor units.

5.2 **Handling of Materials**

5.2.1 Care must be exercised in the handling of all materials. Defective or damaged material must not be installed.

5.2.2 Equipment to load and haul to the job site. All owner-furnished material must be provided. The contractor shall bear the cost of all handling, such as loading, hauling, and unloading.

5.2.3 If framing members (crossarms, bracing, and X-braces) are stored after delivery, they must be arranged with care and placed on blocking at least 1 foot above ground to prevent contact with standing water or the ground. No crossarm must have an unsupported length greater than 20-feet. The blocking must be provided by the contractor and included in the contract's unit prices.

5.2.4 Care must be exercised in handling crossarm assemblies, pole band assemblies, and other factory subassemblies to prevent loss of components for which the contractor is responsible.

5.2.5 Materials or equipment must not be placed where it will be damaged by or cause damage to vehicular traffic, livestock, persons, and property.

5.3 **Special Requirements** (to be completed by the engineer):

81202-11_A03.EPS

Bulletin 1728F-811
Page 6-1

6. STRUCTURE ASSEMBLY

6.1 <u>Reference to Drawings</u>

6.1.1 The contractor shall assemble each structure using the assemblies designated on the plan-and-profile drawings and as shown on the structure and assembly drawings.

6.1.2 Connection details to assemble each structure are referenced on the structure drawings and included with the plans and specifications.

6.2 <u>Structure Framing</u>

6.2.1 The contractor shall frame structures on flat or uniformly sloping terrain located at or near the structure site. Framing on rolling terrain where poles become unsupported should be avoided. If assembly on uniform terrain is not possible, the contractor shall temporarily support the structure components to prevent racking during assembly.

6.2.2 All grid gain teeth must be fully embedded into the wood surface to meet the approval of the engineer. Grid gains must be seated by a combination of tightening assembly bolts and hammering on wood blocking.

6.2.3 All hardware at a connection must be compatible with the fastener diameter. The holes in the hardware must be 1/16 of an inch greater than the fastener diameter, unless otherwise noted. The quantity of square nuts required for a fastener must conform (unless otherwise noted in the List of Materials on the transmission line structure drawings) to ANSI C135.1, <u>Galvanized Steel Bolts and Nuts</u>.

6.2.4 Fasteners must be sized so that they extend not less than 1/2 of an inch nor more than 2-1/2 inches beyond the face of the last nut or locknut. Galvanized bolts must not be cut off unless the engineer allows it for special requirements. Where bolts are not of proper length due to variations in the material, the contractor shall replace the fasteners with ones of proper length at no cost to the borrower.

6.2.5 Spring washers must be installed where specified. The curved portion of the spring washer must be installed horizontally to facilitate inspection of wood shrinkage in the future.

6.2.6 Double crossarms, when installed, must not bow by more than plus or minus 1/2 of an inch per arm. The contractor shall replace spacer fittings, where required, to meet this specification. The additional cost to replace spacers to meet this tolerance must be included in the contractor's unit costs

81202-11_A04.EPS

Bulletin 1728F-811
Page 6-2

for pole top assemblies. Field gaining of poles will not be acceptable.

6.2.7 Pole bands must be sized for the poles upon which they are to be mounted and must fit securely on the poles. Pole bands must be securely tightened around the pole with even spacing between vangs. Where pole bands are not of proper size due to variations in the wood, the contractor shall replace the bands, where required, to met this specification. The additional cost to replace bands must be included in the contractor's unit costs for pole top assemblies.

6.2.8 Pole ground wires must be installed when specified on the plan-and-profile drawings. The downlead must be sufficiently tightened and routed as shown on the respective structure drawings and described on drawing TM-9.

6.2.9 Where X-braces are specified, the lower holes through the pole must be field drilled. These holes must pass through the center of the pole, and must be thoroughly treated with preservative compound approved by the engineer. The heads of machine bolts for mounting the X-braces must be toward the center of the structure.

6.2.10 Guying attachments, where specified, must be oriented as shown on the transmission line structure drawings and as shown on the guying attachment drawings. Pole bands must be installed complete with through bolts, links, grounding clips, and all items indicated on the manufacturer's drawings.

6.2.11 The contractor shall check the end fittings of crossarms, braces, X-braces, and other factory assembled components to see that all factory-installed hardware is secured properly. The cost of retightening factory-installed hardware, if required, must be included in the contractor's unit cost for pole top assemblies.

6.2.12 Assembled structures must be prepared for erection with all items shown in the List of Materials such as conductor assemblies, OHGW assemblies, X-brace assemblies, pole tie assemblies, guying attachments, guying assemblies, and pole ground assemblies.

6.3 **Special Requirements** (to be completed by the engineer):

81202-11_A05.EPS

7. STRUCTURE ERECTION

7.1 Reference to Drawings

7.1.1 The contractor shall check structure locations prior to erecting structures. Structures and specified assemblies must be erected at locations shown on the plan-and-profile drawings.

7.1.2 Tangent structures must be erected as shown on the transmission line structure drawings. The poles or center of H-frames must be placed on the survey centerlines, unless offset left or right of the survey centerlines by the dimension shown on the guying guide drawings or plan-and-profile drawings.

7.1.3 Angle structures and deadend structures must be erected as shown on the structure drawings, guying guide drawings, and plan-and-profile drawings. Angle structures must be placed so that all poles are set on a line perpendicular to the bisector of the line angle. The angle structure must be offset to the left or right of the survey centerline so that all poles are offset by the dimension shown on the guying guide or plan-and-profile drawings.

For deadend structures in which the line angle is 0 degrees, the structures must be placed perpendicular to the survey centerline.

7.2 **Structure Erection**

7.2.1 Tangent structures with single crossarms must be erected with crossarms on alternating sides of the poles. At unusually long spans, the poles must be set so that the crossarms are on the side of the pole away from the long span. At crossings, single crossarms should be attached to the face of the structure away from the crossing.

7.2.2 Care must be taken not to overstress any members or connections when installing structures.

7.2.3 Hardware, bolts, nuts, locknuts, and spring washers must be tight after erection of the structures.

7.3 **Excavation, Setting, and Backfill**

7.3.1 Unless otherwise indicated, all poles must be embedded in soil to a minimum depth of 10 percent of the pole length plus 2 feet and not to exceed 3 inches deeper. Where the ground is sloping, the embedded depth of multiple pole structures with equal pole sizes must be measured on the side of the structure where the ground is lower. (On the high side, greater depth is needed for the purpose of leveling the crossarms.) For multiple pole structures with unequal size poles, poles must be overburied to get the correct ground to phase height.

81202-11_A06.EPS

7.3.2 Pole holes must be a minimum of 8 inches wider than the butt diameter of the pole. When pole bearing plates are used, pole holes must be the minimum diameter necessary for installation of the pole. The excavated hole must be at least as large at the bottom as at the top.

7.3.3 Accumulated water must be removed from the hole prior to setting the pole. Any soil added to level the bottom of the holes must be tamped before the pole is set in the hole.

7.3.4 Pole backfill material must be compactible and suitable for providing a dense supportive soil mass, free of voids, not frozen, and must be approved by the engineer. Where native soil is not suitable for backfill, the contractor shall furnish suitable granular imported material for this purpose which must be paid for at the unit price for granular backfill for poles.

7.3.5 Poles must be set plumb before the backfill is placed; and after placing the backfill, the poles must remain plumb. If the poles are out of plumb, the backfill must be removed and replaced. Plumbing of poles by pushing or pulling the structure must not be permitted.

7.3.6 Backfill must be placed around the pole in layers not exceeding 6 inches in depth, with each layer mechanically tamped before the next layer is added. The backfill must be compacted to a density equal to or greater than that of the surrounding undisturbed soil.

7.3.7 Backfilling and compaction must be done at a rate no faster than one laborer shoveling fill and two others using mechanical or pneumatic tamper.

7.3.8 Native soil must be banked up and tamped around the pole to a height of 6 inches above the natural grade, and must be sloped away from the pole.

7.3.9 After completion of wire stringing, all poles must be reinspected to check that poles remain plumb and the backfill has not settled. The backfill must be retamped at any pole location where the backfill shows settlement or movement. If required by the engineer, the backfill must be completely dug out, the pole readjusted if necessary, and the backfill retamped. This work must be done at no additional cost to borrower.

7.3.10 When approved by the engineer, surplus excavated soil may be carefully spread and leveled on the surface of the ground near the structure and in a manner to minimize damage to the grass areas or other foliage.

7.4 **Special Requirements** (to be completed by the engineer):

81202-11_A07.EPS

9. GROUNDING AND BONDING

9.1 Reference to Drawings: All structures must be grounded as shown on the plan-and-profile drawings and transmission line structure drawings, and subject to the following provisions.

9.2 Structure Grounding

9.2.1 The engineer may require that ground resistance measurements be made for each structure and that additional grounding be added to that already provided by the basic structure grounding assemblies.

9.2.2 Where structure grounding tests are required by the engineer, the contractor shall measure the ground resistance after the structure is erected, but before the overhead ground wire is installed. The method of measuring ground resistance must be subject to the approval of the engineer.

9.2.3 All labor and materials for ground resistance measurements and installation of additional grounding must be provided by the contractor and must be covered by the unit costs for testing and for grounding units.

9.2.4 The contractor shall install counterpoise only after approval of the engineer.

9.3 Bonding of Ground Wire

9.3.1 The pole ground wire must be continuous and not spliced from top of pole to the pole butt grounding assembly. Should damage occur during erection of the structure, the pole ground wire may be spliced with the engineer's approval.

9.3.2 Hardware must be bonded to the pole ground wire as shown on the drawings. The ground wire must clear any unbonded hardware by at least 3 inches.

9.4 Fence and Gate Grounding: Fence and gate grounds must be installed as shown on the drawings. All labor and material required must be furnished by the contractor at the unit prices for fence and gate grounding.

9.5 Special Requirements (to be completed by the engineer):

81202-11_A08.EPS

Appendix B

Framing, Detail, and Guide Drawings for 230kV Tangent H-Frame Wood Structures

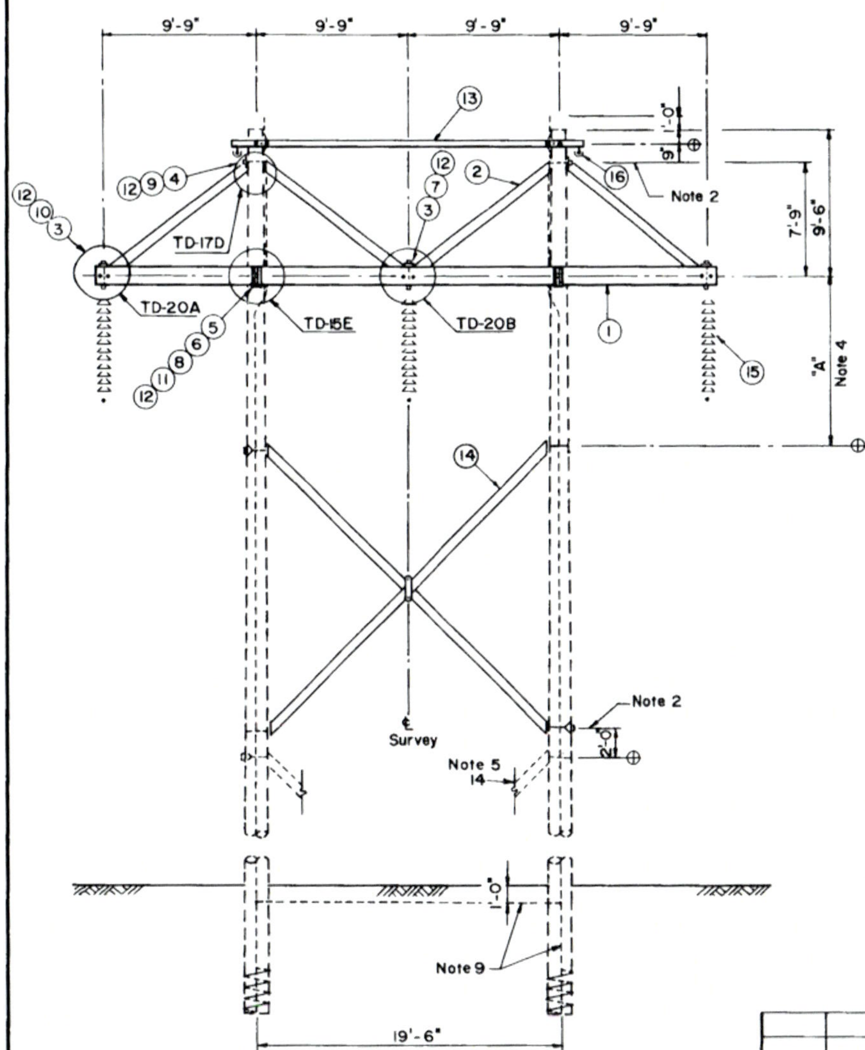

LIST OF MATERIALS

DRG. REF.	QTY.	DESCRIPTION	ITEM	DET.	CODE No.
1	2	X-Arm, 40'-0" (see Note 3), #81		TCD-40	
2	4	Brace, X-Arm, 3-3/8" x 5-3/8" x req'd l.			
3	3	Spacer Assembly, see construction spec.		TM-111	
4	2	Grid Gain, 4" x 4", 15/16" hole			
5	4	Plate, Gain			
6	4	Plate, Ribbed Tie			
7	1	7/8" D.E Bolt, Bent w/2 recessed nuts	gy		
8	2	7/8" Threaded Rod, w/2 nuts			
9	2	7/8" Bolt, Machine, by req'd length			
10	2	7/8" Bolt, Bent			
11	2	Washer, Spring, 15/16" hole			
12	10	7/8" Locknut, MF Type			
13	1	OHGW Support Assembly	-	TM-7D	
14	1	X-Brace Assembly	vx	TM-110C	
15	3	INSULATOR ASSEMBLY, TANGENT	-	TM-____	
16	2	OHGW ASSEMBLY, TANGENT		TM-4____	

Notes:

1. Double X-Arms shall be shipped with factory assembled hardware.
2. Field drilled holes shall be thoroughly treated.
3. Engineer to specify X-Arm size by type number. X-Arm may be 3-5/8" x 9-3/8 solid sawn (X-Arm #81) or 5-1/8" x 7-1/2" laminate (X-Arm #83).
4. Dimension "A" shall be as shown on the pole framing drawing.
5. Structure TH-230X is the same as TH-230 with one additional X-Brace.
6. For other assembly requirements, see REA specification T-8.
7. For strength limitations of OHGW support assembly, see drawing TM-7D.
8. Drawing TE-2 gives guidance to subassembly alternatives.
9. The following materials are to be specified on plan and profile drawings and staking sheets: POLES, POLE GROUNDING ASSEMBLIES, AND ANY POLE FOUNDATION UNITS.

Reissued 03/98

NO	REVISION	DATE

TRANSMISSION LINE STRUCTURE

TANGENT H FRAME

(230 kV MAXIMUM)

SCALE

TH-230

81202-11_A09.EPS

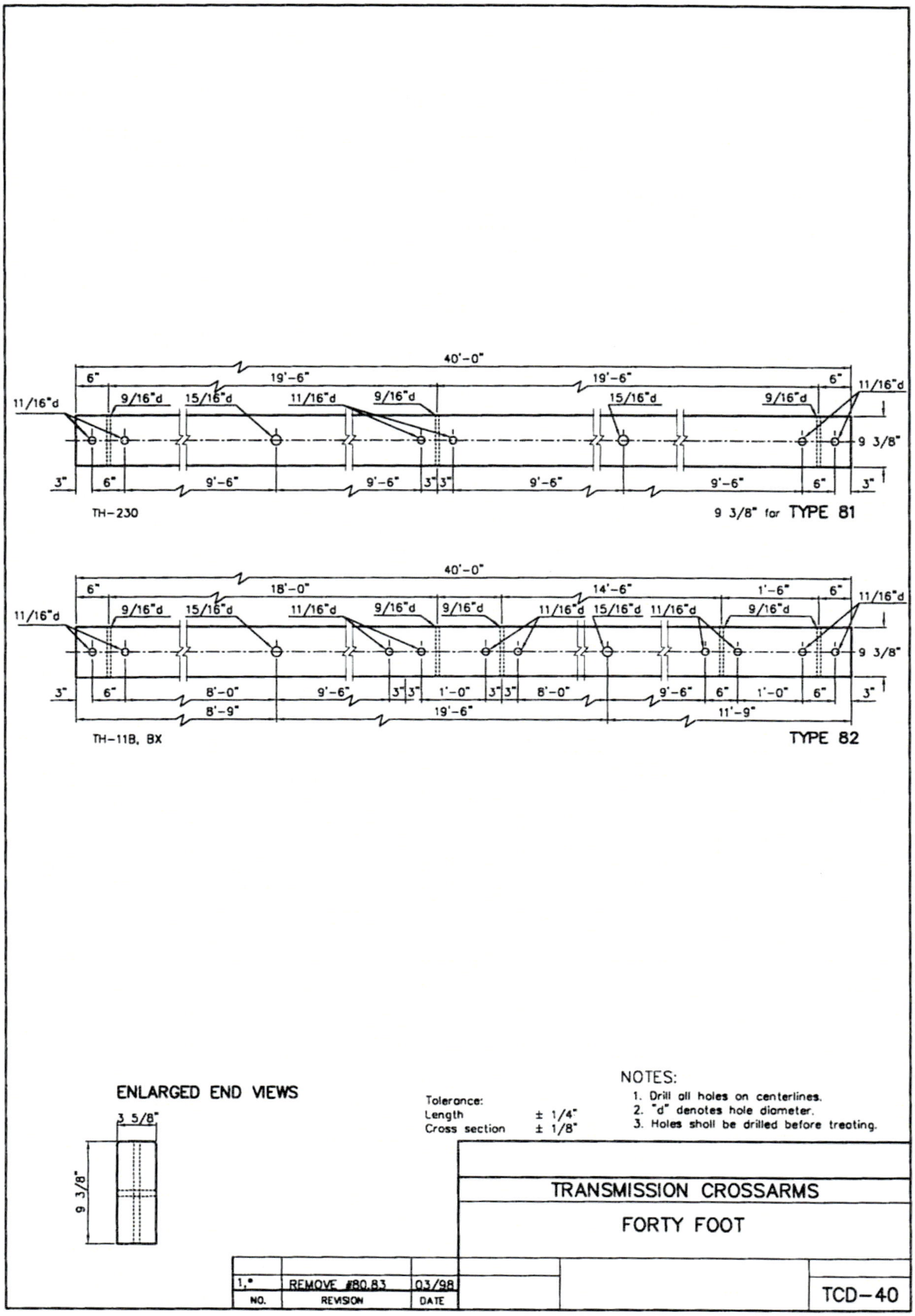

81202-11_A10.EPS

DWG. REF.	TM-7 A	B	C	D		LIST OF MATERIALS DESCRIPTION	ITEM	DET.	CODE No.
1	2	2	2	2		7/8" Bolt, Machine, by req'd length	c		
2	2	2	2	2		5/8" U-Bolt, w/4 nuts			
3	2	2	2	2		Chain Link, 5/8" x 2 1/4"	br		
4	2	2	2	2		Washer, Curved, 4"sqx1/4",15/16" hole	d		
5	2	2	2	2		Washer, Spring, 15/16" hole	aw		
6	2	2	2	2		Clamp, Groundwire, 1/2"	dp		
7	2	2	2	2		7/8" Locknut, MF Type	ek		
8	1	-	-	-		Angle, Tie, 3 1/2 x 3 x 1/4 x 15'-0"	-	A	
9	-	1	-	-		Angle, Tie, 3 1/2 x 3 x 1/4 x 18'-0"	-	B	
10	-	-	1	-		Angle, Tie, 5 x 3 x 1/4 x 18'-0"	-	C	
11	-	-	-	1		Angle, Tie, 5 x 3 x 1/4 x 22'-0"	-	D	
12									
13									

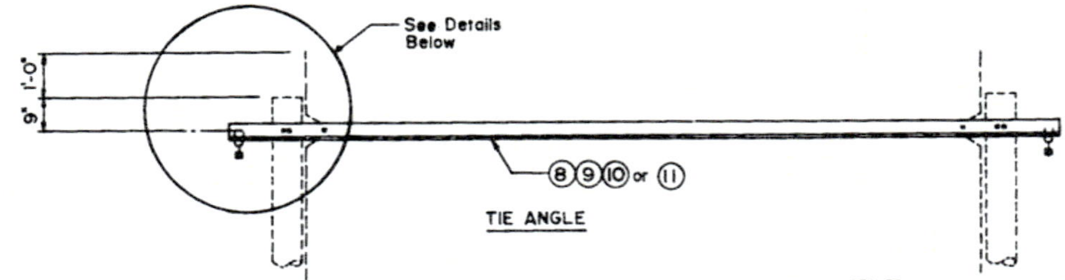

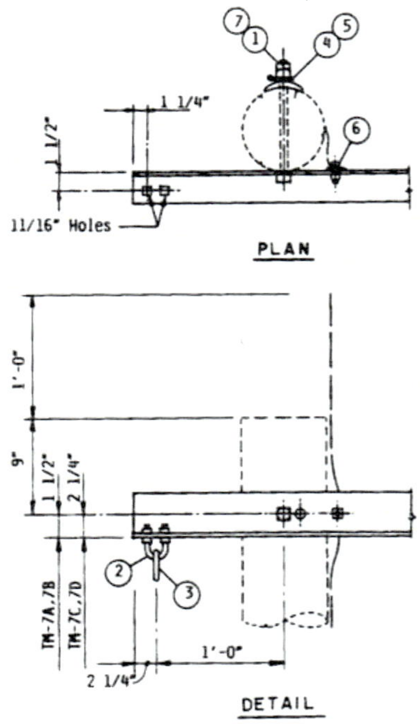

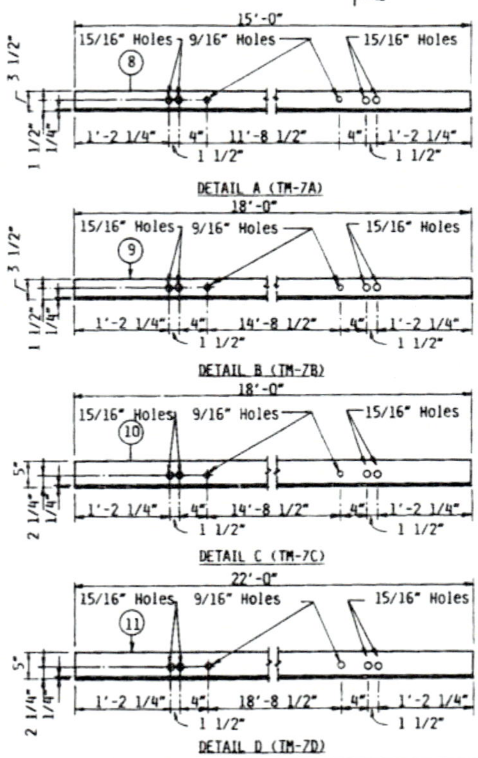

NOTES:

1. The strength limitations of the OHGW Support Assemblies are:

	Allowable Vertical Load	Maximum Vertical Load
TM-7A, 7B	1200 lbs.	1900 lbs.
TM-7C, 7D	2300 lbs.	3700 lbs.

2. Tie angle steel shall conform to ASTM A36. Angle shall be hot dip galvanized per ASTM A123.
3. Substitute TM-7C for TM-7B when specified on TH-10 or TH-11B.

OHGW SUPPORT ASSEMBLY

STEEL TIE ANGLE

No.	REVISION	DATE	SCALE:		TM-7

Reissued 03/98

81202-11_A11.EPS

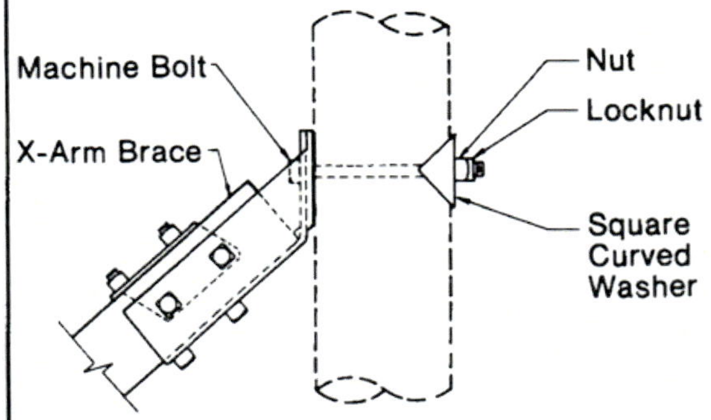

TD-17A

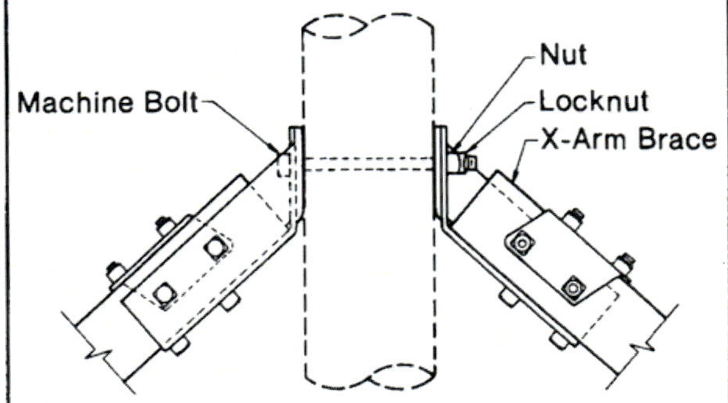

TD-17B

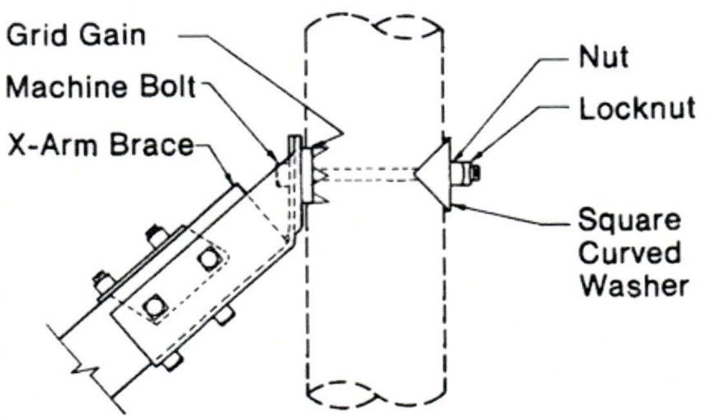

TD-17C

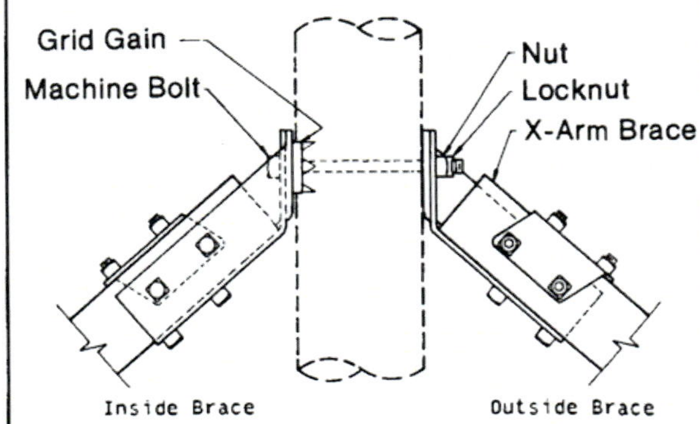

TD-17D

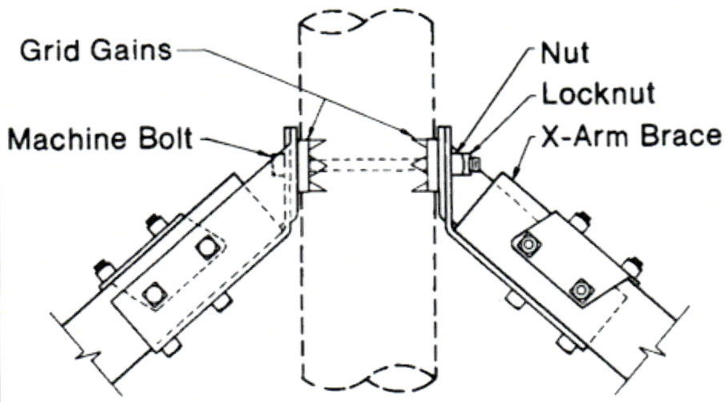

TD-17F

DETAILS

BRACE TO POLE CONNECTIONS

NO.	REVISION	DATE	SCALE:	SHEET 1 of 2
				TD-17

Reissued 03/98

81202-11_A12.EPS

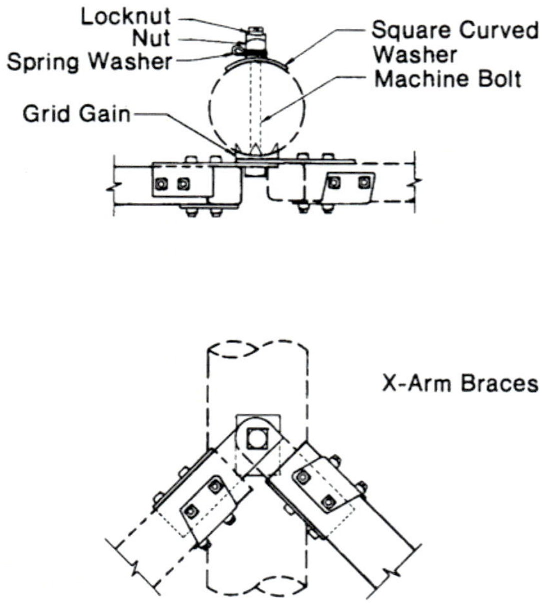

TD-17G

(SINGLE OR DOUBLE BRACE)

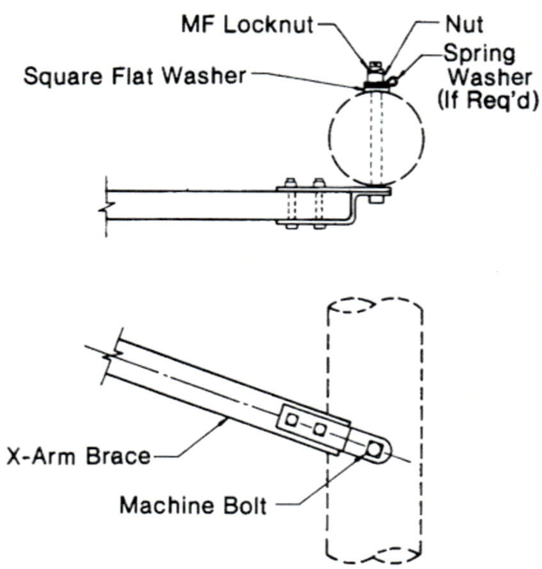

TD-17I

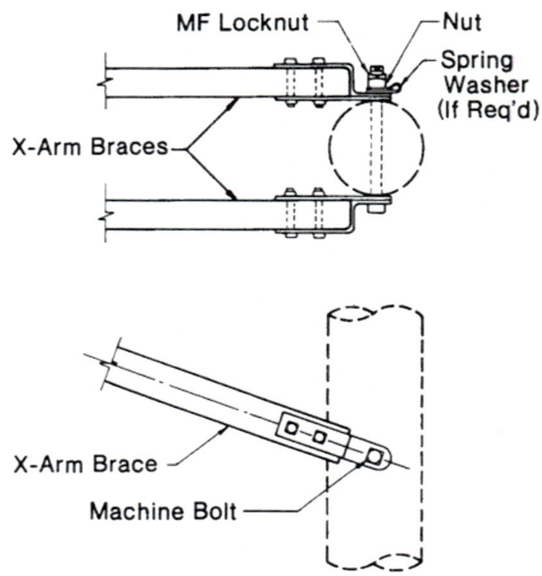

TD-17J

DETAILS

BRACE TO POLE CONNECTIONS

SHEET 2 of 2

TD-17

Reissued 03/98

81202-11_A13.EPS

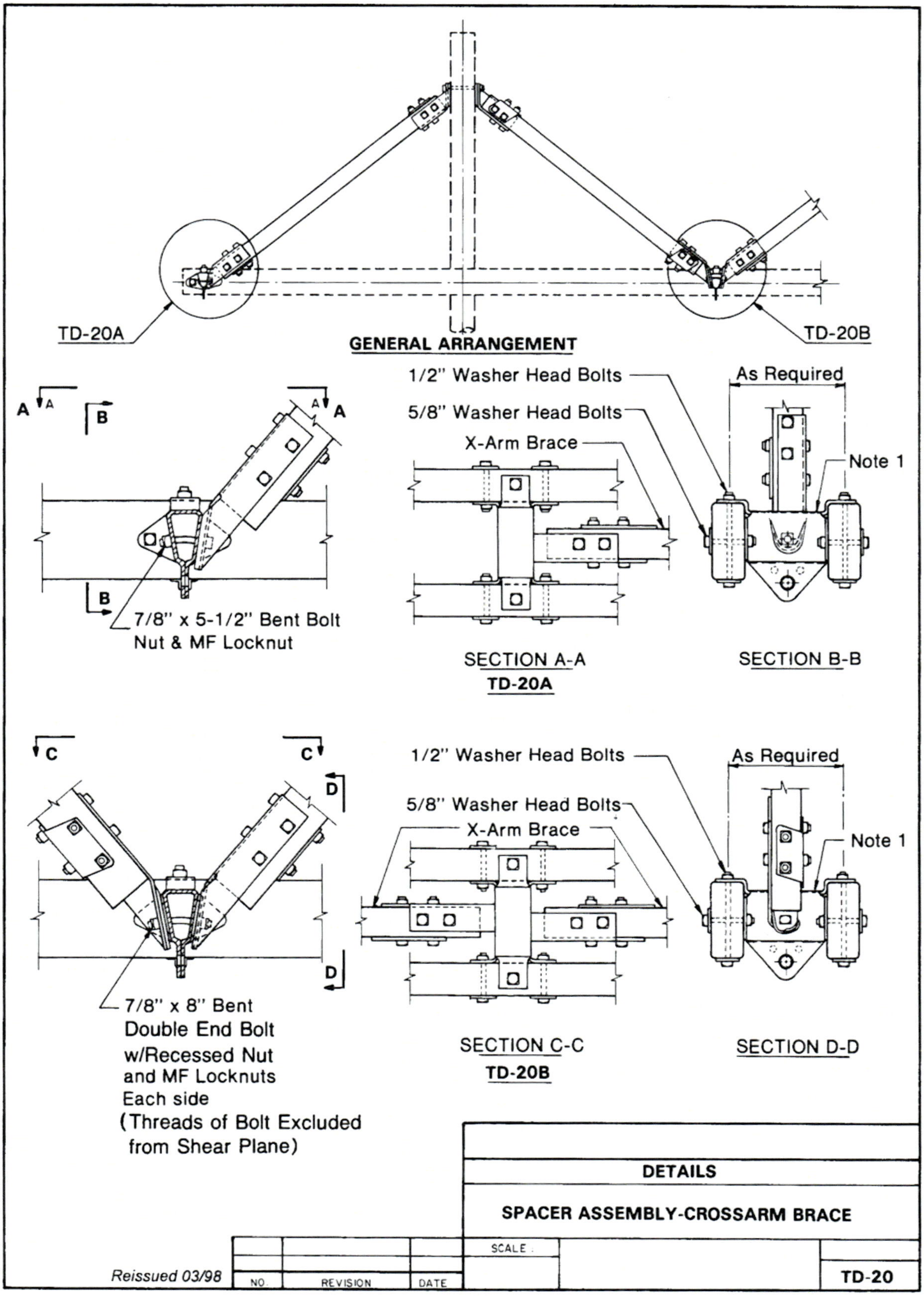

81202-11_A14.EPS

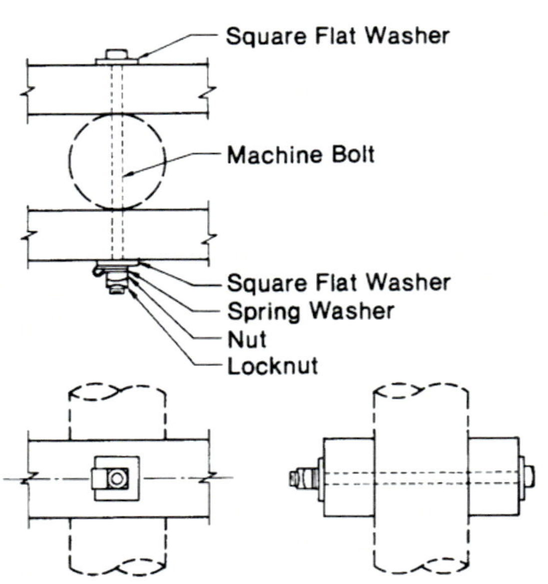

TD-15A

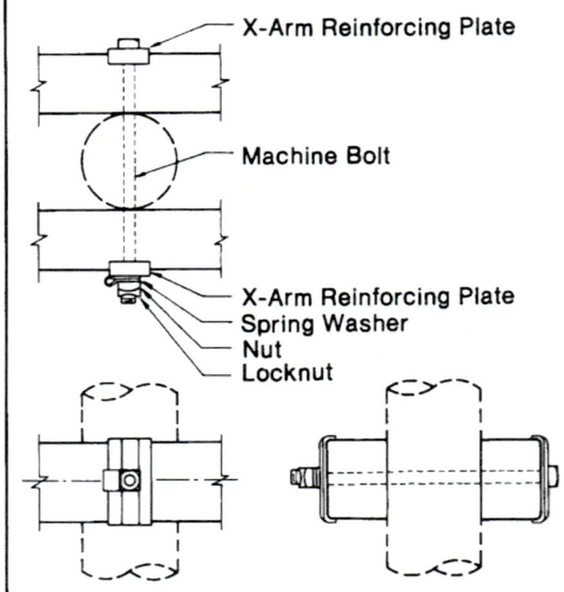

TD-15B

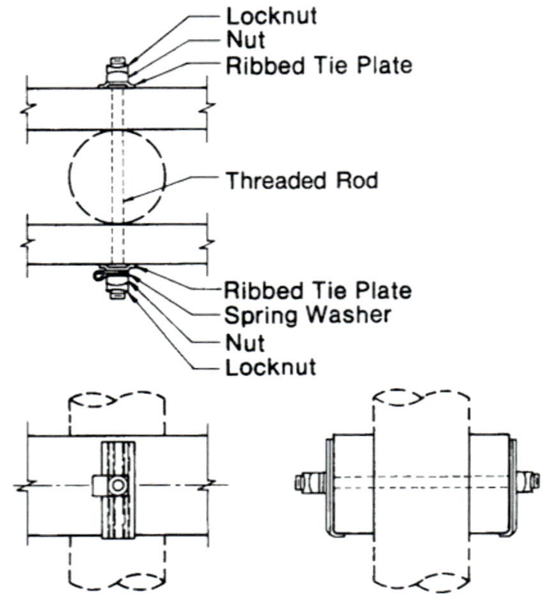

TD-15C

TD-15D

DETAILS

DOUBLE CROSSARM TO POLE

NO.	REVISION	DATE	SCALE:	SHEET 1 of 2
				TD-15

Reissued 03/98

81202-11_A15.EPS

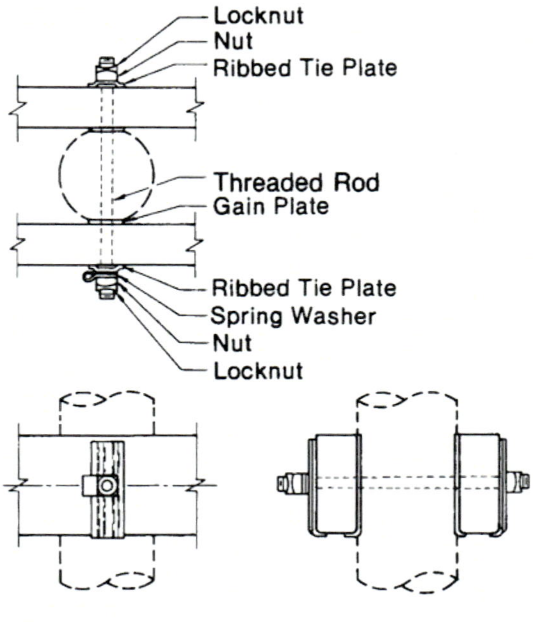

TD-15E

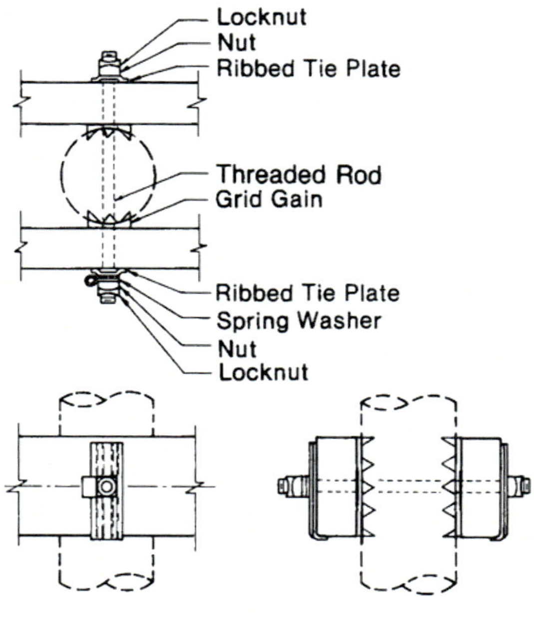

TD-15F

DETAILS

DOUBLE CROSSARM TO POLE

Reissued 03/98

NO.	REVISION	DATE	SCALE:		SHEET 2 of 2
					TD-15

81202-11_A16.EPS

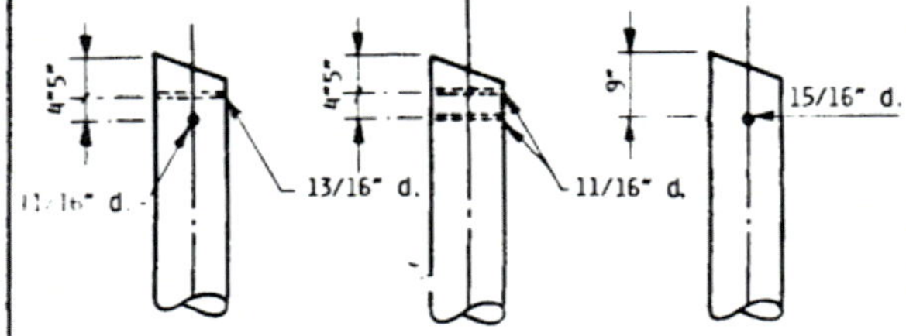

Single Bolt
TM-6A

Double Bolt
TM-6B

Tie Angle
TM-7

OHGW SUPPORT ASSEMBLY

DETAIL A

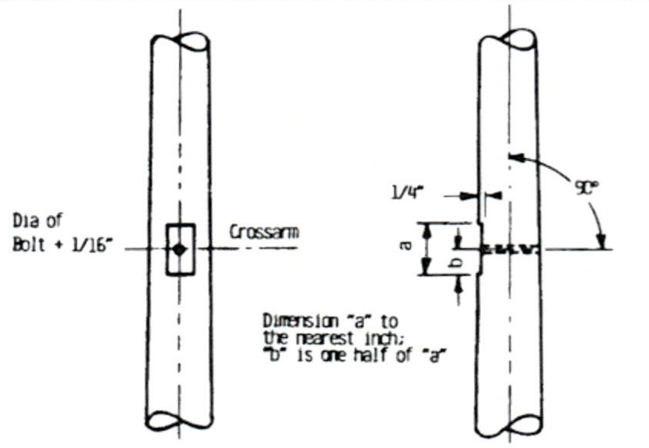

POLE GAIN FOR MOUNTING
CROSSARMS (NO SPACERS)

DETAIL B

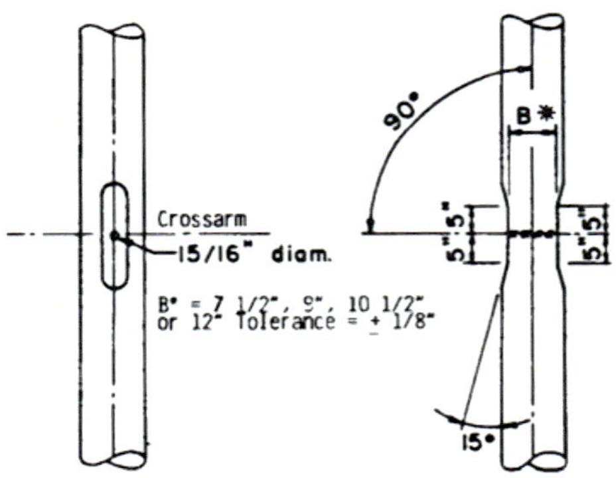

POLE GAIN FOR MOUNTING
CROSSARMS (FIXED SPACERS)

DETAIL C

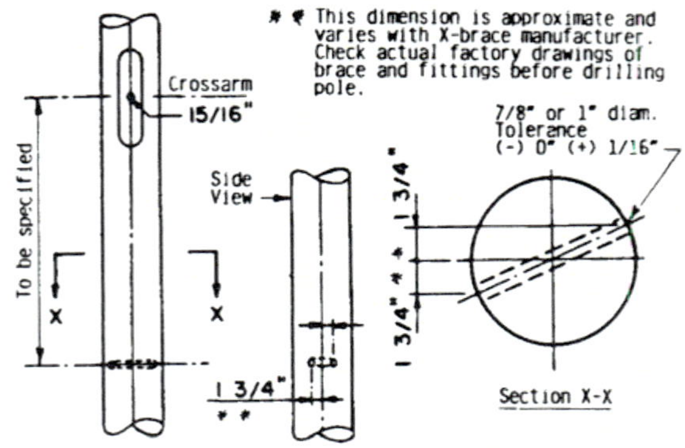

POLE DRILLING FOR
CROSSBRACE (TM 110A, B, C)

DETAIL D

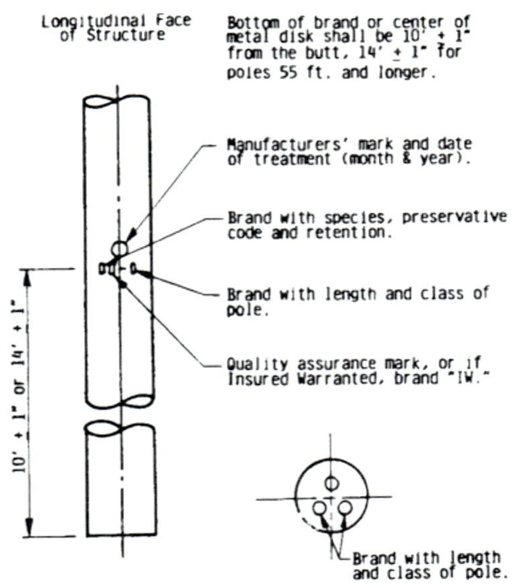

BRANDING

DETAIL E

Reissued 03/98

NOTES:

1. Poles and treatment shall conform to REA Specification DT-5C latest revision.
2. All poles treated full length must be bored (except where otherwise specified), roofed, and gained before treatment.
3. Refer to the construction specification to determine if roofs are to be flat or at an angle of 15°.
4. Gains shall be flat with their plane at right angles to bolt holes.
5. All field drilled holes shall be thoroughly treated.
6. Crossarms may be bowed ± 1/2".

TYPICAL NOTES

TRANSMISSION POLES

POLE FRAMING GUIDE DETAILS

Aug., 1986		**TPF-1**

81202-11_A17.EPS

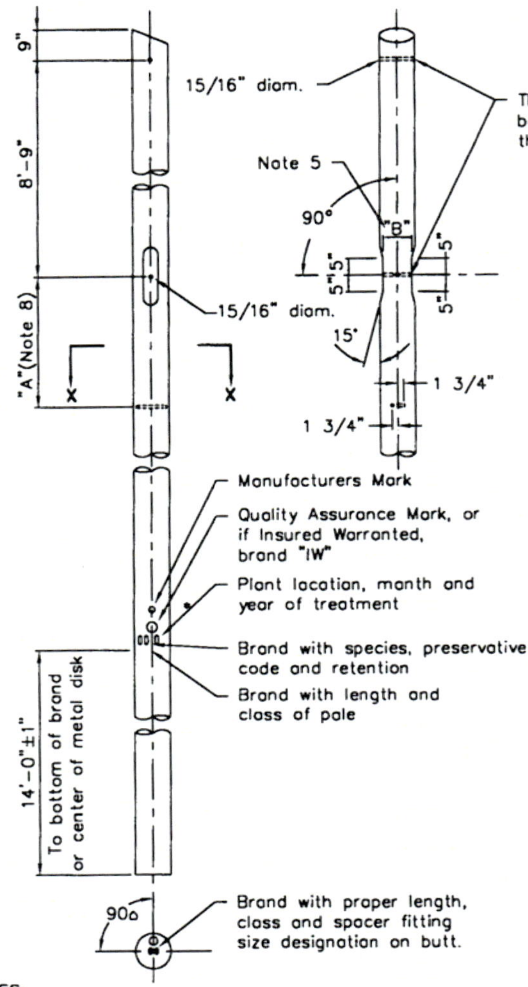

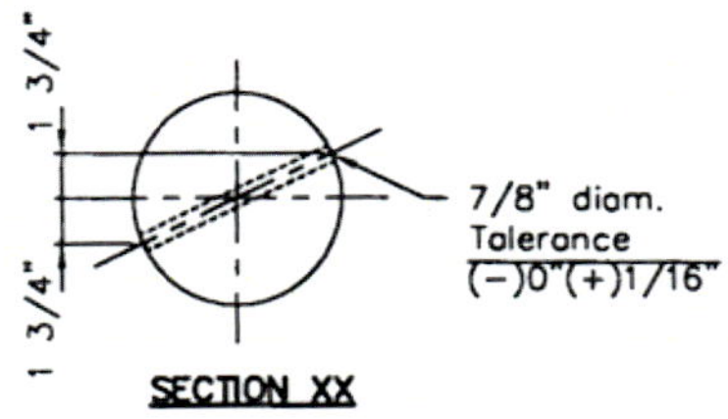

POLE HEIGHT FEET	DIMENSION "A" 230 kv
60	
65	
70	
75	
80	
85	
90	
95	
100	
105	

NOTES:

1. • Poles and treatment shall conform to RUS Specifications on Wood Poles, Stubs, and Anchor Logs.
2. All poles treated full length must be bored (except where otherwise specified), roofed, and gained before treatment.
3. Refer to the construction specification to determine if roofs are to be flat or at an angle of 15°.
4. Gains are to be flat with their planes at right angles to the bolt holes.
5. Separation between double gains shall conform to Dimension "B" in the table (tolerance ±1/8).
6. • Refer to the construction specifications as to gaining of the poles when using adjustable spacers.
7. Crossarms may be bowed ±1/2".
8. Dimension "A" shall conform to the table (Engineer to insert proper dimensions).
9. The hole at Section XX may be bored prior to treatment or field drilled at contractor's option unless otherwise specified.
10. All field drilled holes in full length treated poles shall be pressure treated.
11. Poles in each structure shall be matched in size, strength and straightness.
12. This dimension is approximate and varies with X-Brace manufacturer. Check actual factory drawings of brace and fittings before drilling poles.

Circumference of Pole at Point of Gain Inches	Dimension "B" Inches	Spacer Fitting Size	Spacer Fitting Designation
21.0 – 21.9	6.5	7 1/2	A
22.0 – 23.5	7.0	7 1/2	B
23.6 – 25.1	7.5	7 1/2	C
25.2 – 26.7	8.0	9	D
26.8 – 28.2	8.5	9	E
28.3 – 29.8	9.0	9	F
29.9 – 31.4	9.5	10 1/2	G
31.5 – 33.0	10.0	10 1/2	H
33.1 – 34.5	10.5	10 1/2	I
34.6 – 36.2	11.0	12	J
36.3 – 37.7	11.5	12	K

TRANSMISSION POLES

STRUCTURE TH-230

Rev 03/98•

TPF-50

81202-11_A18.EPS

LIST OF MATERIALS

DWG. REF.	TM- 101	TM- 102A	TM- –	TM- 103	DESCRIPTION	ITEM	DET.	CODE No.
1	cu.yd	cu.yd	–	cu.yd	Imported Backfill (Note 1 & 2)	–	–	
2	–	–	–	1/2yd	Concrete, 2000 psi. (Note 9)	–		
3	–	2	–	–	Pole Bearing Plates	fs		
4	–	1	–	–	7/8" Bolt, Machine by req'd length	c		
5	–	–	–	–	5/8" Bolt, Machine by req'd length	c		
6	–	1	–	–	7/8" Clamp, Groundwire +1 nut (Note 5)	dp		
7	–	1	–	–	7/8" Locknut, MF Type	ek		

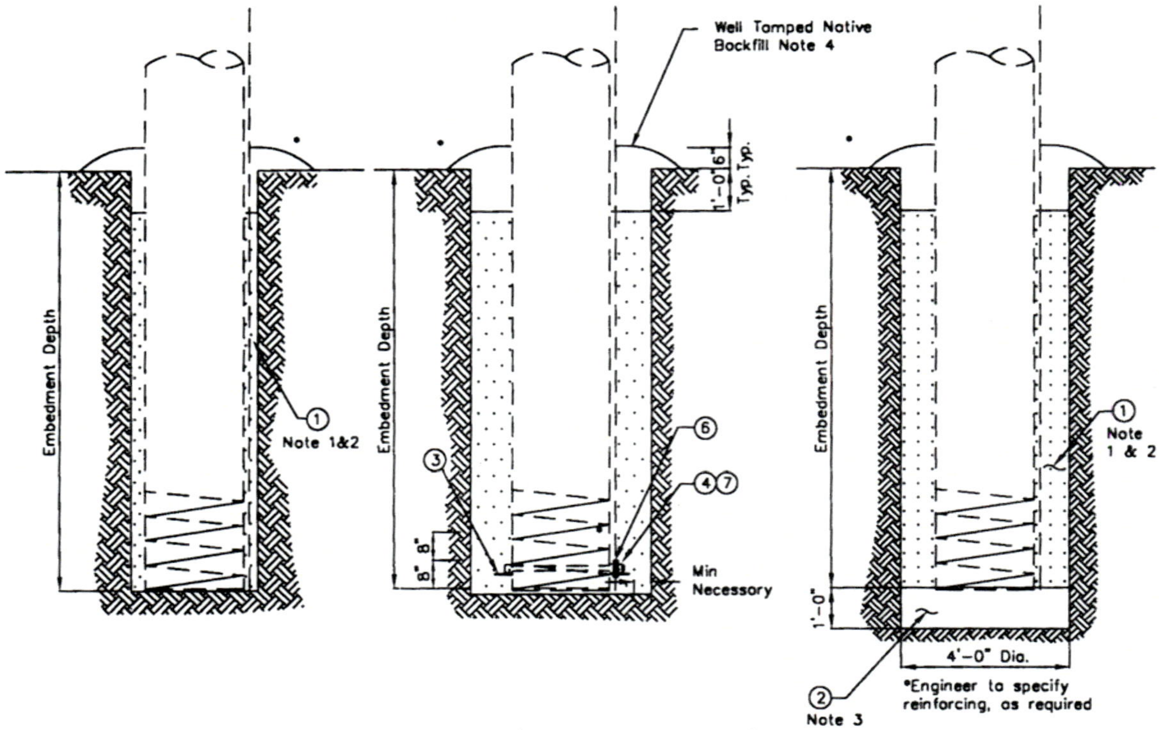

TM–101 • TM–102A (2 BEARING PLATES) TM–103

NOTES:

1. Gradation for imported backfill is shown in the table. The aggregate shall be well mixed in a stock pile, and pneumatically placed and tamped in 6–inch lifts.

2. The specification for aggregate given in the table is minimum. In areas where smaller fines are available at comparable cost, it is recommended that the engineer specify smaller fines.

3. Concrete shall be 2000 psi high early strength concrete. It shall be deposited in a manner which will prevent segregation of the aggregate and cement. Care should be taken to avoid getting dirt mixed with the concrete. The pole shall not be set until the concrete bearing base has attained sufficient strength to support the pole. During placement of pole, avoid bumping concrete base.

4. Well tamped native backfill should be placed around the pole as shown.

5. All buried metals must be consistent with good corrosion prevention practices. Similar or compatible metals should be used for bearing plates, pole ground wire and ground connectors.

GRADATION FOR IMPORTED GRANULAR BACKFILL	
SIZE OF GRAVEL (OR) CRUSHED STONE	SIZE OF MESH (in)
100% By Weight to pass 1" screen	1.00
60% – 90% By Weight to pass 1/2" screen	0.500
40% – 60% By Weight to pass no. 4 screen	0.187
25% – 50% By Weight to pass no. 8 screen	0.0937
20% – 40% By Weight to pass no. 16 screen	0.0469
15% – 30% By Weight to pass no. 40 screen	0.0165

FOUNDATION UNITS

POLE STABILITY, BEARING, & UPLIFT FOUNDATIONS

NO.	REVISION	DATE
1,*	TM101,102A,103,dr.6	03/98

TM–101,102,103

81202-11_A19.EPS

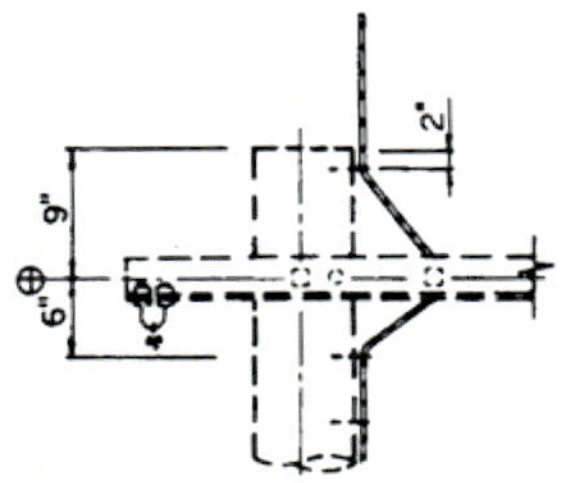

TM-9				LIST OF MATERIALS			
DRG. REF.	A	B	C	DESCRIPTION	ITEM	DET.	CODE No.
1	ft.	-	ft.	5/16" Wire, soft annealed iron	cj		
2	lb.	-	-	2" x 1/2" Staples, galvanized steel	al		
3	-	ft.	-	#6 or #4 Wire AWG, copper or copperclad	cj		
4	-	lb.	-	1-1/2" x 3/8" Staples, copperclad	al		
5	-	-	2	Clamp, U-Bolt	u		

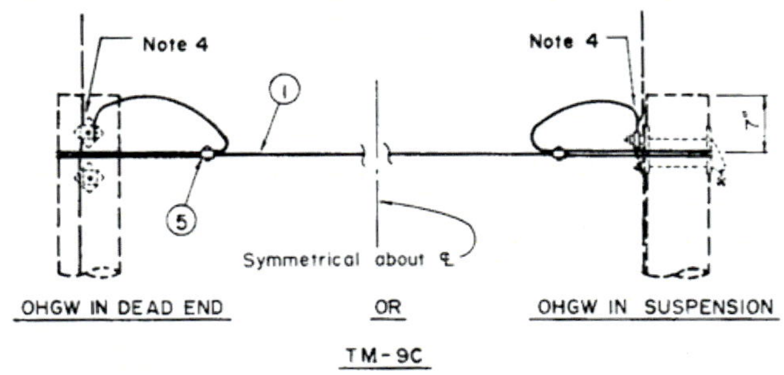

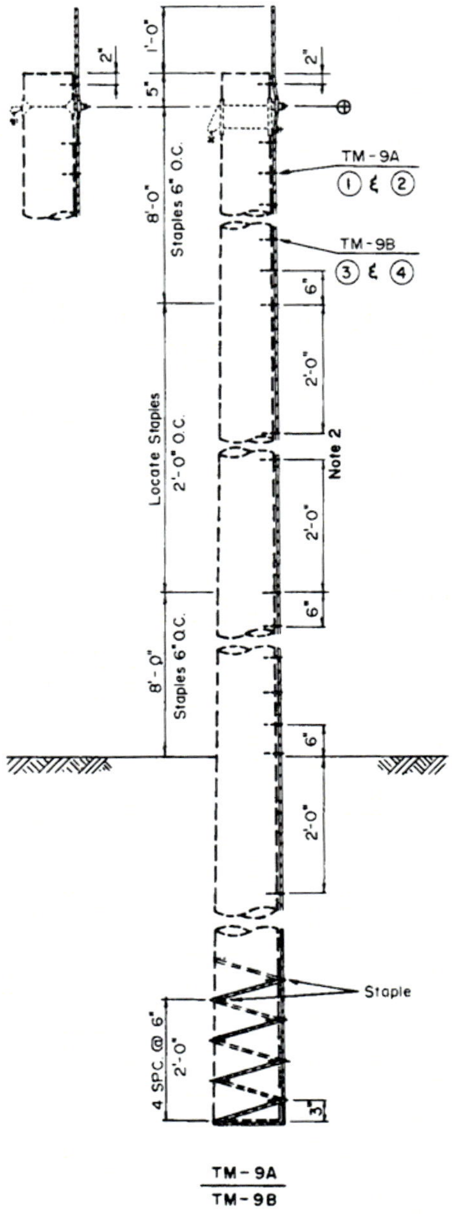

TM-9A
TM-9B

NOTES:

1. Orientate the downlead as shown on the transmission line structure drawings. On structures with X-Arms, the downlead shall be located and stapled so that it does not come in contact with the X-Arm or the through bolts.

2. Staple the downlead to the pole leaving one foot of wire projecting above the poles. Staples on the downlead shall be 2 feet apart, except for a distance of 8 feet above the ground and 8 feet from the top of the pole where they shall be 6 inches apart.

3. Carry the downlead under the bottom of the pole, and then wrap the downlead four turns back up the pole. Staple the groundwire to the pole at each wire crossing and on the opposite sides of the pole.

4. Terminate the tie wire for TM-9C at the groundwire clamp.

GROUNDING ASSEMBLIES

POLE GROUND AND BUTT WRAP

No.	REVISION	DATE			
			Aug., 1986		1 OF 2 — **TM-9**

Reissued 03/98

81202-11_A20.EPS

	TM - 9		LIST OF MATERIALS			
DRG. REF.	D, F, H, J, L,	E, G, I, K, M,	DESCRIPTION	ITEM	DET.	CODE No.
1	as req'd	-	Pole Ground & Butt Wrap	-	TM-9A	
2	ft.	-	5/16" Wire, soft annealed iron	cj		
3	as req'd	-	2" x 1/2" Staples, galvanized steel	al		
4	as req'd	-	Connector, compression, for item 2	p		
5	as req'd	-	Connector, galvanized steel			
6	-	as req'd	Pole Ground & Butt Wrap	-	TM-9B	
7	-	ft.	#6 or #4 Wire AWG, copper or copperclad	cj		
8	-	as req'd	1-1/2" x 3/8" Staples, copperclad	al		
9	-	as req'd	Connector, compression, for item 7	p		
10	-	as req'd	Connector, copper (or) bronze			

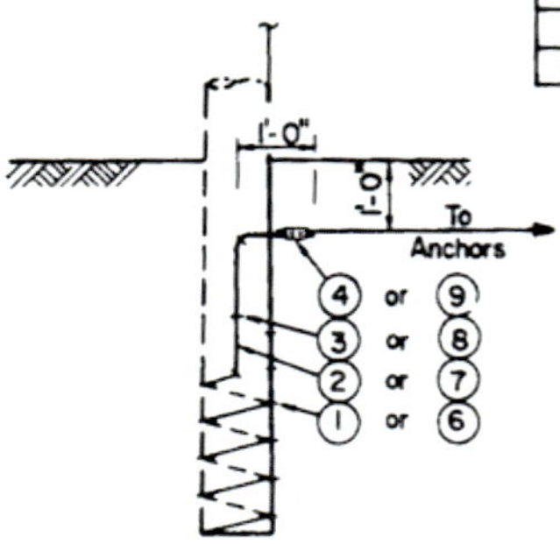

TM-9D( )
TM-9E( )
Note 4

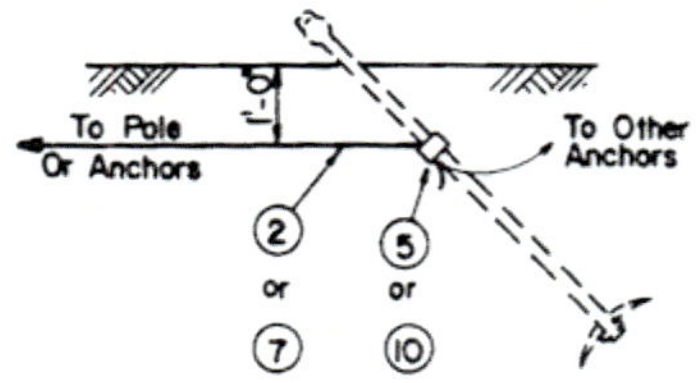

DETAIL A

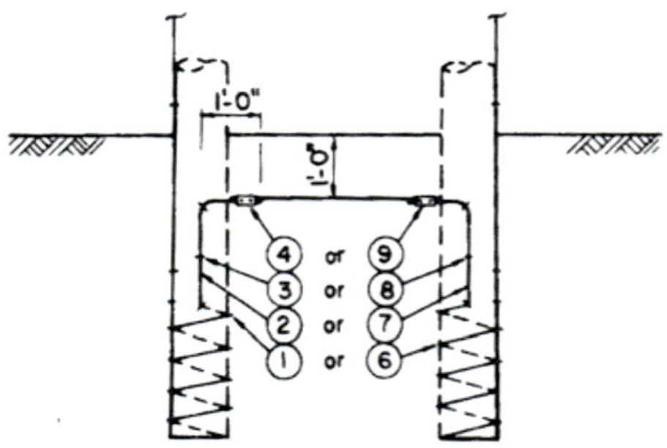

TM-9F( )
TM-9G( )

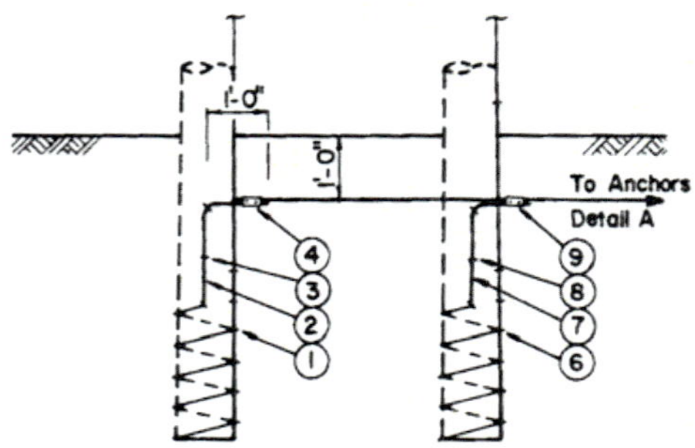

TM-9H( )
TM-9I( )

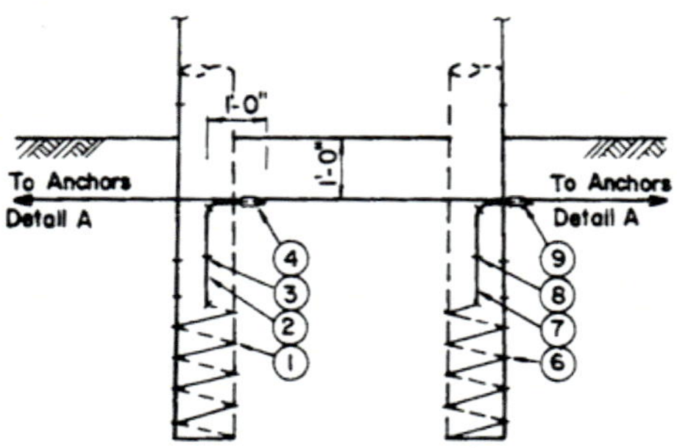

TM-9J( )
TM-9K( )

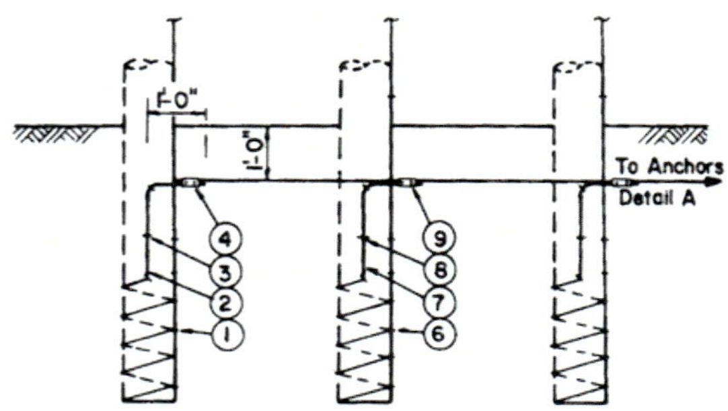

TM-9L( )
TM-9M( )

Notes:

1. For other installation details, refer to sheet 1 of 2, TM-9.
2. Grounding materials must be consistent with good corrosion prevention practices. When galvanized hardware and 5/16" iron wire are used, anchor rods shall be galvanized steel. When copper, copperclad or bronze hardware are used, anchor rods shall be copperclad.
3. Bid items on this drawing include material to top of pole in accordance with sheet 1 of 2, TM-9.
4. ( ) indicates number of anchors.

GROUNDING ASSEMBLIES

MULTIPLE POLE AND ANCHOR GROUNDING

NO	REVISION	DATE	Aug., 1986	2 OF 2
				TM-9

Reissued 03/98

81202-11_A21.EPS

Additional Resources

This module presents thorough resources for task training. The following resource material is suggested for further study.

Electric Power Transmission Systems, Second Edition. Upper Saddle River, NJ: Prentice Hall.

Figure Credits

Photograph courtesy of American Electric Power, Module opener, Figures 9, 19, and 30

Topaz Publications, Inc., Figures 1, 3–8, 10–17, 18 (B, D), 22–24, 36, 37, and 46

KT Power, Inc., Figures 18 (A, C), 25, 31, 34, and 43

USDA Rural Development, Figure 20, Appendix A and B

Bonneville Power Administration, Figure 26

California Public Utilities Commission, Figures 27, 33, and 35

Dryden & LaRue, Inc., Figure 28

Roterra Screw Piling Ltd., Figure 29

Hubbell Power Systems, Figures 38–41

Photo courtesy of Thomas & Betts Corporation, all rights reserved., Figure 42

Photograph provided by AEMC® Instruments, Figure 44

NCCER CURRICULA — USER UPDATE

NCCER makes every effort to keep its textbooks up-to-date and free of technical errors. We appreciate your help in this process. If you find an error, a typographical mistake, or an inaccuracy in NCCER's curricula, please fill out this form (or a photocopy), or complete the online form at **www.nccer.org/olf**. Be sure to include the exact module ID number, page number, a detailed description, and your recommended correction. Your input will be brought to the attention of the Authoring Team. Thank you for your assistance.

Instructors – If you have an idea for improving this textbook, or have found that additional materials were necessary to teach this module effectively, please let us know so that we may present your suggestions to the Authoring Team.

NCCER Product Development and Revision

13614 Progress Blvd., Alachua, FL 32615

Email: curriculum@nccer.org
Online: www.nccer.org/olf

❑ Trainee Guide ❑ AIG ❑ Exam ❑ PowerPoints Other ______________________

Craft / Level: Copyright Date:

Module ID Number / Title:

Section Number(s):

Description:

Recommended Correction:

Your Name:

Address:

Email: Phone:

81203-11

Transmission Equipment Installation

Module Four

Trainees with successful module completions may be eligible for credentialing through NCCER's National Registry. To learn more, go to **www.nccer.org** or contact us at **1.888.622.3720**. Our website has information on the latest product releases and training, as well as online versions of our *Cornerstone* newsletter and Pearson's product catalog.

Your feedback is welcome. You may email your comments to **curriculum@nccer.org,** send general comments and inquiries to **info@nccer.org**, or fill in the User Update form at the back of this module.

V.1 12/11

81203-11
Transmission Equipment Installation

Objectives

When you have completed this module, you will be able to do the following:

1. Explain tower and pole rescue procedures.
2. Describe the effects of electrostatic and electromagnetic induction.
3. Install temporary grounding to create an equal potential zone (EPZ).
4. Prepare a transmission structure for conductors by installing:
   - Insulators
   - Yoke plates
   - Stringing blocks
   - Pilot line
   - Pulling cable
5. Use Kellems grips socks and double Kellems grips to install:
   - Phase conductors
   - Overhead shield conductors and optical ground wire
6. Operate single- and multiple-conductor pulling and tensioning machines.
7. Splice conductors using compression sleeves.
8. Identify and install different types of connectors including compression and AMPACT™ types.
9. Sag an installed conductor using:
   - Sag scope or transit
   - Dynamometer
   - Return wave timing
10. Complete the installation of a conductor by:
    - Dead-ending the conductor
    - Clipping-in the conductor and removing stringing block
    - Installing dampers and spacers
    - Installing warning lights and day markers

Performance Tasks

Under the supervision of your instructor, you should be able to do the following:

1. Install temporary grounding to create an equal potential zone (EPZ).
2. Prepare a tower for conductors by installing:
   - Insulators
   - Yoke plates
   - Stringing blocks
   - Pilot line
   - Pulling cable
3. Use Kellems grips and double Kellems grips to install:
   - Phase conductors
   - Overhead shield conductors and optical ground wire
4. Operate single- and multiple-conductor pulling and tensioning machines.
5. Splice conductors using compression sleeves.
6. Identify and install different types of connectors including:
   - Compression
   - AMPACT™

continued

Performance Tasks (*continued*)

7. Sag an installed conductor using:
   - Sag scope or transit
   - Dynamometer
   - Return wave timing
8. Complete the installation of a conductor by:
   - Dead-ending the conductor
   - Clipping-in the conductor and removing stringing block
   - Installing dampers and spacers
   - Installing warning lights and day markers

Trade Terms

Armor rod
Articulated running board
Bullwheel tensioner
Compression sleeve
Corona ring
Double Kellems grip
Drum-style pulling machine
Dynamometer
Electromagnetic induction
Electrostatic induction
Equal potential zone
Full-tension splice
Ground mat
Kellems grip
Minimum approach distance
Optical ground wire (OPGW)
Pilot line
Rescue stretcher
Running ground
Sag scope
Spacer cart
Step potential
Stringing block
Suspension clamp
Tension stringing method
Touch potential
Yoke plate

Industry Recognized Credentials

If you're training through an NCCER-accredited sponsor you may be eligible for credentials from NCCER's Registry. The ID number for this module is 81203-11. Note that this module may have been used in other NCCER curricula and may apply to other level completions. Contact NCCER's Registry at 888.622.3720 or go to www.nccer.org for more information.

Contents

Topics to be presented in this module include:

Figures and Tables

Figures and Tables (*continued*)

1.0.0 Introduction

Transmission lines carry the power produced at generating facilities to regional and local distribution substations. Transmission line work presents unique hazards due to the voltage levels and working heights. Voltages carried on transmission lines can range from 115 kV to 750 kV, and transmission structures can be 100 feet or more above the ground. Once a transmission structure is erected, power line workers must install hardware and string conductors between the structures. Because of the potentials and the working heights, installing and maintaining transmission lines is inherently dangerous, so the use of safe work practices is critical. This module explains how to safely install transmission line conductors and related equipment.

2.0.0 Safety

All power line workers are required to wear personal protective equipment (PPE) to prevent on-the-job injuries.

- *Clothing* – Clothing should be made of natural fibers, such as cotton or wool, and treated to be arc-resistant and fire-retardant. Avoid clothing made of certain synthetic fibers, such as rayon, nylon, and polyester, which could burn or melt and cling to skin. Pant legs should extend over the top of shoes or boots. Never wear shorts on the job. Shirts should be long-sleeved to protect the skin from the sun, to minimize injuries caused by wood splinters, and to provide protection from chemicals used to treat the wood against rot. On most job sites, workers are required to wear a bright-colored shirt or vest to increase their visibility.
- *Footwear* – Only safety-toe leather shoes or boots should be worn on the job. High-top boots have the added advantage of greater support for the ankles. Many companies require dielectrically rated boots or overshoes. Never wear sandals or canvas-type shoes on the job.
- *Eye protection* – Safety glasses with side shields in accordance with *ANSI Standard Z87.1* must be worn at all times while on the job. If a power tool such as a chain saw is used on the job, a full face-shield may be necessary in addition to the safety glasses.
- *Hearing protection* – If power tools or noisy machinery, such as a chain saw, trencher, or air compressor, are used on the job, approved earmuffs or earplugs are generally required.
- *Dust mask* – If a wood pole is cut or drilled, a dust mask must be worn. The preservatives in the resulting sawdust may be harmful if inhaled.
- *Gloves* – Working on wood poles used in aerial distribution systems can be rough on hands. Leather work gloves provide protection from rope burns, wood splinters, and other hazards.
- *Rubber gloves and sleeves* – Transmission line workers often work near energized lines or equipment. Rubber gloves and sleeves protect against electrical shock. Use an air and water test on the gloves and roll the sleeves in each direction to inspect for damage. Inspect all other PPE and tools according to the manufacturers' recommendations to ensure they are safe to use.
- *Safety helmet/hard hat/hard cap* – An approved *ANSI Standard Z89.1 Class E* safety helmet made of a non-conducting plastic material, such as polypropylene or polycarbonate, must be worn at all times while on the job site. The helmet protects the head from injuries caused by bumps and falling objects, and it provides protection from the sun.

In addition to wearing the proper PPE, follow these safety precautions when working with transmission equipment:

- Avoid installing or removing aerial equipment in wet or bad weather, except in emergencies.
- If applicable, use protective covers, line hose, rubber blankets, and shields around conductors and other components when working near energized lines.
- Before starting any job, inspect and test rubber gloves and sleeves using the water and/or air method, and ensure that all PPE and tools are safe to use.
- Only use tools and equipment that are approved for transmission line use.

On Site

Transmission Structure Types

Transmission structures come in a variety of shapes and sizes. Two wood or metal poles are often installed as an H-frame. Lattice steel towers (LSTs) are widely used. Tubular steel poles (TSPs) are becoming more popular.

- Use fall protection equipment when working at heights above 6 feet (4 feet for unqualified climbers).
- If an insulated lift bucket is used for bare hand work, test the insulating arm for excessive leakage current each day before it is used. A non-barehand bucket requires an annual boom leakage test.
- Have a qualified observer at all job sites to ensure that **minimum approach distances** to energized equipment are observed, safe work practices are being followed, and correct PPE and tools are being used.

> **NOTE**
> A qualified observer is a member of the work crew with the primary duty of ensuring that clearances to energized equipment are maintained, safe work practices are followed, and appropriate PPE and protective shields are being used. That person is familiar with all aspects of installation and construction of distribution systems and is trained in first aid. While performing the duties of a qualified observer, that person cannot perform any other tasks.

2.1.0 Electric Shock Hazards

Power line workers are protected from electric shocks by two methods. One method is isolation from energized surfaces by distance or by working with insulated tools such as a hot stick. The second method is by providing a work zone in which all surfaces that can be touched are at the same electrical potential.

Electric shock hazards that exist around transmission structures have three main sources:

- Energized conductors
- **Electrostatic induction**
- **Electromagnetic induction**

Work around energized conductors should be avoided whenever possible. However, when you are required to work in the vicinity of an energized conductor, wear insulating PPE or use other insulating methods. Maintain minimum approach distances to energized surfaces when performing the required task, as dictated by company policy. Some typical minimum approach distances are listed in *Table 1.*

Electrostatic induction is a process by which a charged object can induce a charge in a nearby object. Objects near or under a high-voltage power line can become charged in this manner. If a worker touches a charged object, that charge can pass through the worker's body to ground. Utilities restrict the type of activity that can take place under a power line right-of-way. Crops may be allowed, but metal irrigation pipes must be grounded. Metal fences are discouraged. Machines used to harvest crops, and trucks used to service the power lines should not extend above 14 feet in height. Vehicles should not be parked under power lines unless they are properly grounded to bleed off any induced charge.

> **NOTE**
> Utilities do not allow buildings to be constructed within a power line right-of-way. Buildings located outside of, but near a right-of-way should have all metal components grounded to bleed off any induced charge.

Electromagnetic induction is a process by which a conductor that carries alternating current can induce a current in a nearby conductor (*Figure 1*). Transformers operate by electromagnetic induction. The changing electromagnetic field surrounding an energized line that runs parallel to a de-energized line can induce a current in the de-energized line. If a worker touches that de-energized line and that line has not been properly grounded, he or she can receive a shock. Potentials that occur in objects as the result of electrostatic or electromagnetic induction are managed by proper grounding.

2.2.0 Reducing Electric Shock Hazards

One of the best ways to prevent electric shocks is to provide a work zone in which all surfaces that can be touched are at the same electrical potential. That work zone is known as an **equal potential zone**. Not only is this a safe practice, it is mandated by law. *OSHA regulation 1910.269(n)(3)* states, "Temporary protective grounds shall be placed at such

Table 1 Typical Minimum Approach Distances

Voltage Range (kV)	Approach Distance*	
	Low Volts	High Volts
72.6 to 121	3'-5"	4'-3"
230 to 242	5'-3"	7'-6"
345 to 362	8'-6"	12'-6"
500 to 550	11'-3"	18'-1"
765 to 800	14'-11"	26'

*May vary. Always maintain distances specified by the local utility.

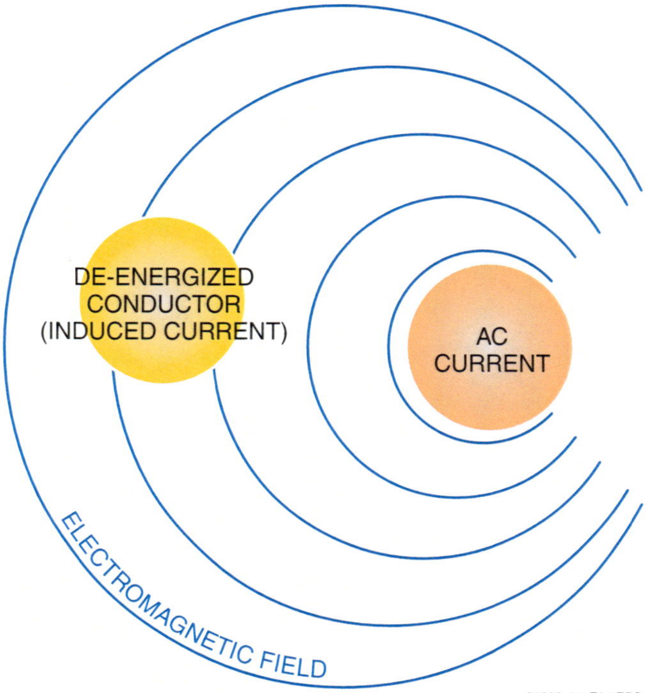

Figure 1 Electromagnetic induction.

locations and arranged in such a manner as to prevent each employee from being exposed to hazardous differences in electrical potential."

It is a difference in potential that causes current to flow and an electric shock to result. In *Figure 2*, the personnel platform and conductor must be tied together to a common ground in order to create an equal potential zone.

Many grounding practices that were considered safe in the past do not always create a safe work zone. *Figure 3* shows a typical grounding practice where all phase conductors on either side of the work area are grounded. However, the hookup shown can actually create an unsafe condition. The de-energized section of line with protective grounds on either side can become the secondary of a transformer. The energized line that runs parallel to the de-energized line can induce a current into the parallel line that is the work area (*Figure 4*). A worker within that area who touches a conductor and ground would receive a shock.

> **NOTE**
>
> Establishing an equal potential zone is as much an art as it is a science. Experts agree that creating an equal potential zone is not a perfect system, but it currently provides the best protection for workers.

When an energized grounded object such as a metal tower leg touches the ground, it energizes the soil at the point of contact. Voltage is highest at the

Figure 2 Equal potential zone.

point of contact with the soil and decreases as the distance from the point of contact increases (*Figure 5*). A person standing with both feet together near the contact point is safe because both feet are at the same potential. However, if one step is taken, one foot can be at one potential and the other foot at a different potential. This potential difference, called

On Site

Birds on a Wire

Have you ever wondered why birds perched on a hot conductor do not get electrocuted? The reason is simple. Both their feet are at the same potential. Without a difference of potential, there can be no current flow.

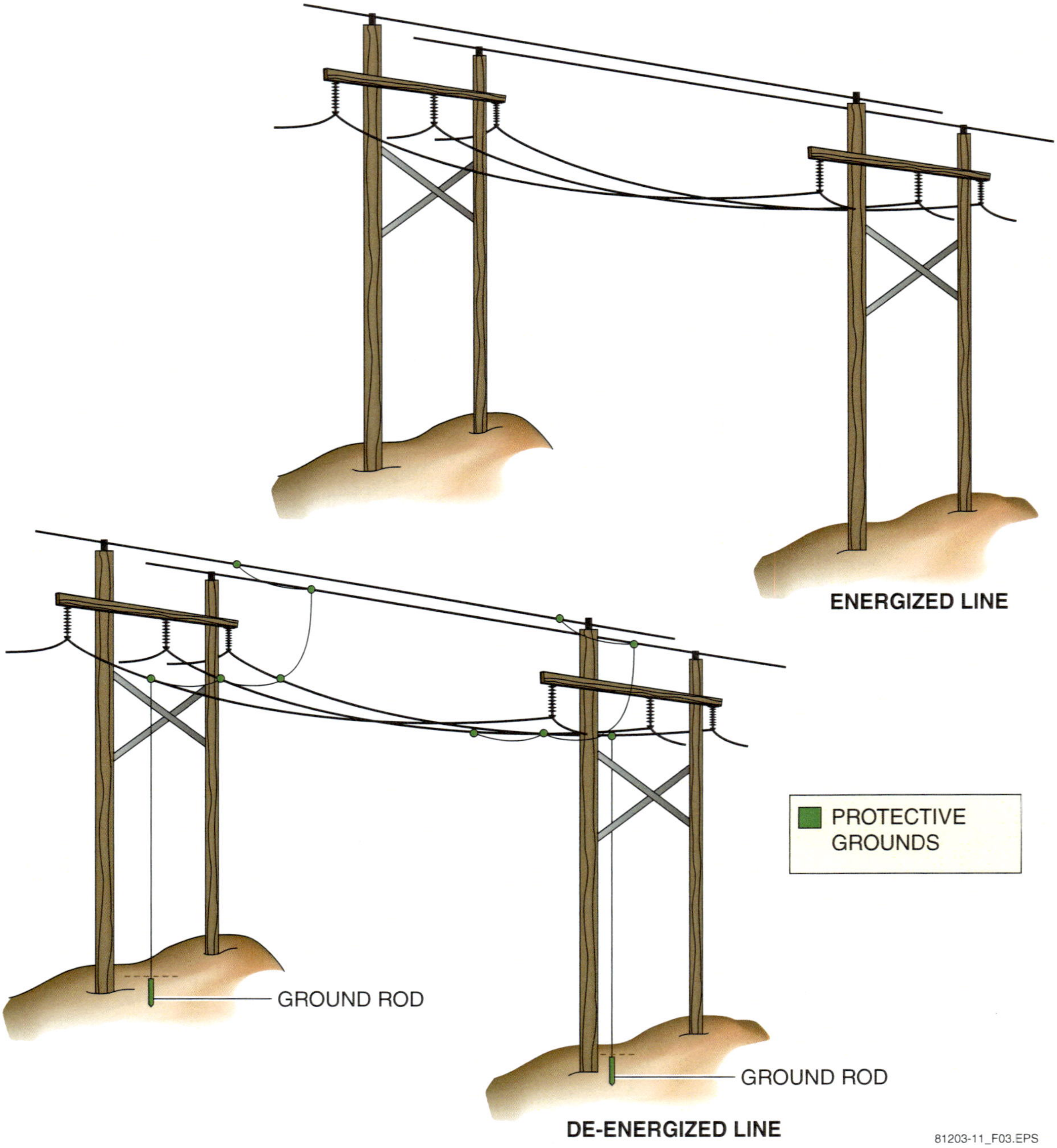

Figure 3 Typical grounding practice.

the **step potential**, can cause a shock. If you ever get in this situation, *do not* walk away to remove yourself. Instead, shuffle your feet, or hop with both feet together to exit the danger zone. *Figure 6* shows feet in a way that can create a step potential. Energized soil around an energized grounded object can also create a **touch potential** hazard. If a person standing on energized soil touches the energized grounded object, a potential is created between that person and the object (*Figure 7*). Step

On Site

Deadly Static Electricity

Air moving over an object can cause a static charge to build up. On low-humidity days, wind blowing over a de-energized power line can build up a large static charge if that line is not grounded. Line workers have been killed by the static charge on a de-energized line.

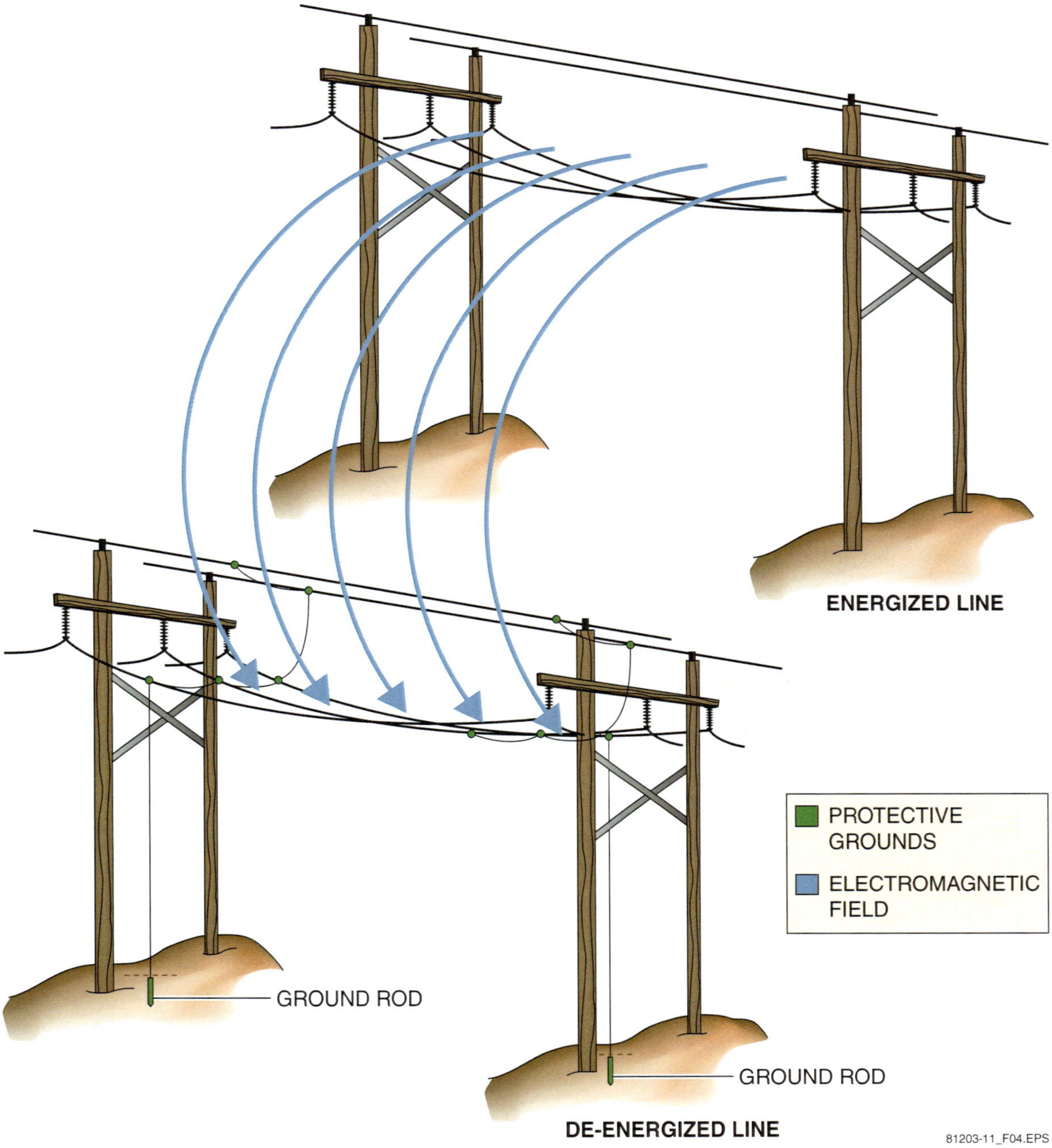

Figure 4 Induced voltage in protected area.

and touch potential hazards can be eliminated by placing a **ground mat** (*Figure 8*) next to and connected to the energized grounded object. The ground mat sets up an equal potential zone (*Figure 9*) by placing the object and the mat at the same potential. The worker can then safely stand next to and/or touch the energized grounded object without getting a shock.

> **NOTE**
> Soil conditions (soil type, dampness, etc.) at the point where an energized grounded object contacts the soil can affect the area that the step or touch potential occupies.

> **WARNING!**
> Stepping off a ground mat exposes a worker to step and touch potential hazards. When stepping off a ground mat, always move away from the energized grounded conductor.

3.0.0 Transmission Structure Rescue Procedures

Workers who become injured while working on a transmission structure must be removed in a timely and safe manner. Co-workers may take part in some or all of the rescue activities. Before attempting a rescue, a worker needs to do the following:

- Find out the nature and extent of the injury.
- Determine what needs to be done to safely remove the worker from the transmission structure.
- Ensure your own safety before attempting a rescue.

A worker's injuries can range from minor to severe. If the worker is conscious and not bleeding, your response will be different than if he or she is unconscious or is not breathing. If you are a distance away from the injured worker, make your way to his or her location without putting your own safety at risk. Once you are with the other worker, determine the extent of the injury. If he or she is unconscious and not breathing, immediately call 911 and start mouth-to-mouth resuscitation. An unconscious person may be upside down or in an awkward position. You must get them upright or in a better position.

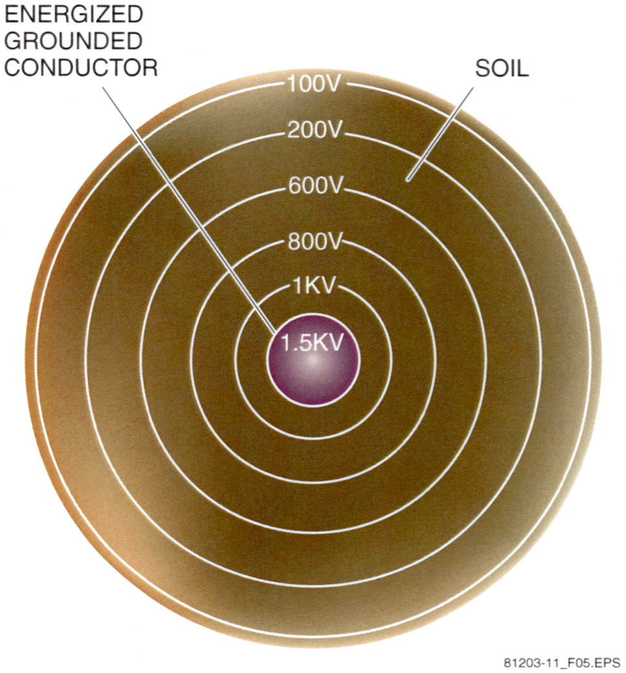

Figure 5 Energized soil around an energized grounded conductor.

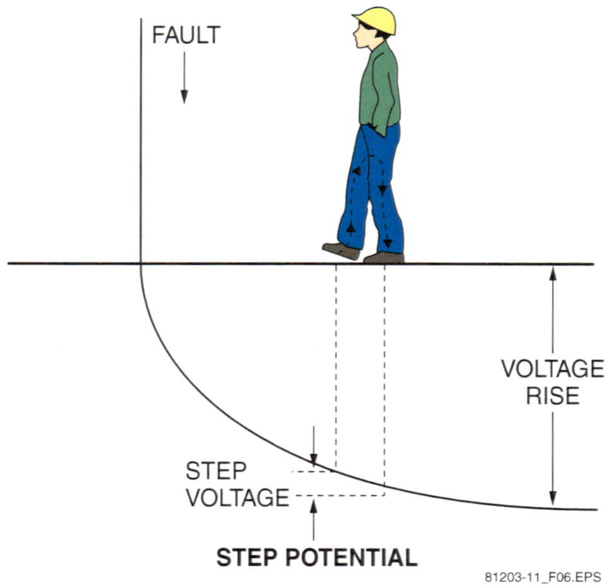

Figure 6 Step potential hazard.

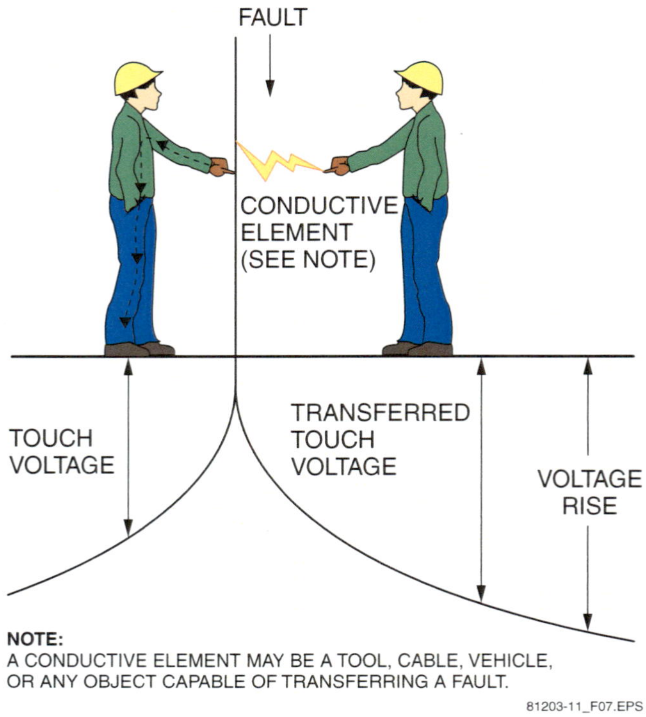

Figure 7 Touch potential hazard.

Figure 8 Ground mat.

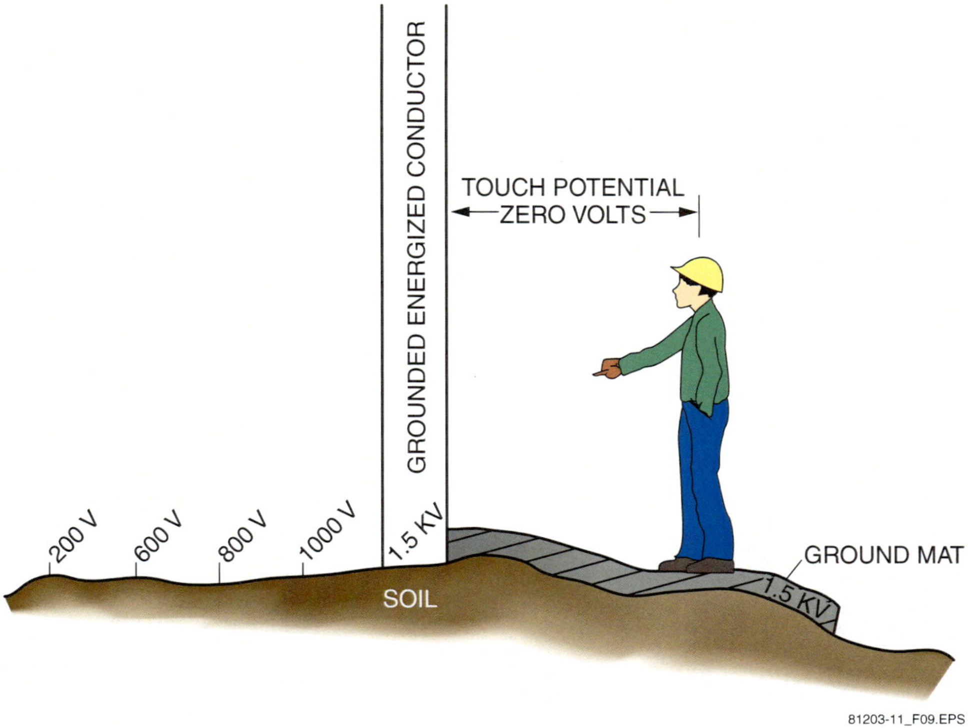

Figure 9 Ground mat equal potential zone.

NOTE

All power line workers should know how to perform CPR and first aid. Classes are available that teach these skills.

WARNING!

If the injured worker has received an electric shock, first make sure that all conductors near the worker are de-energized. For your safety and the safety of the injured worker, install temporary grounds and establish an equal potential zone in the area where the rescue is to take place.

If the injury is minor, you may only be required to assist the worker to leave the transmission structure by normal means. If the worker is bleeding or has broken bones, use first aid to control the bleeding and/or make the person more comfortable. Once you have determined the extent of the injury and have given first aid, you can see what must be done to safely remove the injured worker from the transmission structure. If the worker is conscious and the injuries are not too severe, he or she can be lowered with a line attached to their fall protection harness. If the worker is unconscious or has severe injuries, a **rescue stretcher** (*Figure 10*) may be required to lower the worker. The rescue line is securely attached to the transmission structure and then to the fall protection harness or other lowering device. Tension is put on the rescue line to relieve tension on the lanyard attached to the fall protection harness. The lanyard is then released from the harness to allow the worker to be lowered to the ground with the rescue line. Once on the ground, paramedics can render additional aid and/or transport the worker to a hospital.

4.0.0 Preparing a Transmission Structure for Conductors

After transmission structures are erected, hardware must be installed to allow conductors to be strung between transmission structures. On each transmission structure, insulators may be installed first. A **stringing block** is then attached to each insulator. A **pilot line** is then strung through the stringing blocks along the line of transmission structures.

On Site

Rescue Professionals

Power line workers receive transmission structure rescue training. However, there may be times when the conditions of the rescue are beyond the ability of power line workers. In those cases, others specially trained in high rescues must be called.

81203-11_F10.EPS

Figure 10 Rescue stretcher.

4.1.0 Insulators

Insulators may be made of glass, ceramic, porcelain, or polymer material. Higher voltage transmission lines use cap and pin insulators made up of disc-shaped insulator bells that are strung together and connected with clevis pin or ball and socket links. Different line voltages are accommodated by changing the number of insulator units in a string. This unitized design also allows individual insulator discs to be replaced. Insulators can be installed in a number of different ways. The type of transmission structure often determines the way insulators are installed. For example, suspended insulators are widely used on wood H-frame and lattice steel towers. Horizontal and braced insulators can be found on poles. *Figure 11* shows examples of the kinds of suspended insulators used on transmission structures. This module focuses on insulators installed on lattice steel towers (*Figure 12*).

> **NOTE**
> Polymer insulators are widely used today because of their lighter weight and ability to resist damage.

81203-11_F11.EPS

Figure 11 Suspension-type polymer insulators.

4.1.1 Installing Insulators

Insulators can be heavy, so they are often installed on the transmission structure while it is on the ground and before it is lifted into place with a crane (*Figure 13*). An insulator stack is made up of a number of discs or bell assemblies that are strung together to achieve the level of insulation required. The insulators should be secured so they do not swing or become damaged as the transmission structure is lifted. At other times, the insulators are installed after the transmission structure is erected. A single insulator may be suspended vertically, or two insulators can be suspended at 45-degree angles from vertical to support one or more phase conductors. The insulator(s) is secured to the transmission structure with an anchor shackle. The lower end of the insulator is connected directly to the **suspension clamp** if one insulator is supporting one phase conductor (*Figure 14A*). If two insulators are used, both lower ends are connected to a **yoke plate** (*Figure 14B*). One or more suspension clamps can be attached to the yoke plate. Different types of end fittings are available for insulators (*Figure 15*).

> **NOTE**
> Suspension clamps are attached to the yoke plate after the phase conductors are strung.

81203-11_F12.EPS

Figure 12 Insulators installed on a tower.

81203-11_F13.EPS

Figure 13 Insulators installed on the ground.

High voltage can cause the air around a power line to ionize. This causes a visible glow called corona. Corona can result in radio-frequency noise, audible noise, and it can damage insulators. A device called a **corona ring** (*Figure 16*) can be clamped around insulators to reduce the effects of corona.

4.2.0 Installing Stringing Blocks

A stringing block (*Figure 17*), also known as a traveler, may be installed below each suspended insulator. They are attached to the insulator or to the yoke plate. Stringing blocks are available for single or multiple (bundled) conductors. The stringing block allows conductors to be pulled smoothly between the transmission structures.

4.3.0 Installing a Pilot Line

Phase conductors are heavy. It would be very hard to directly lift these heavy conductors up the transmission structure for stringing. To string conductors, a lightweight line, called a pilot line or P-line, is first threaded through the stringing block. The pilot line is used to pull through a heavier pulling rope or cable, which is then used to pull the heavy conductor through the stringing block. If the structure is very high, the pilot line may be strung using a helicopter.

A

B

81203-11_F14.EPS

Figure 14 Insulator connections.

81203-11_F16.EPS

Figure 16 Corona rings on insulators.

5.0.0 Installing Conductors

Conductors must be installed with care to prevent damage to the conductors. The complete installation procedure includes the following:

- String the conductors
- Tension, sag, and dead-end conductor to the transmission structures
- Remove stringing blocks
- Clip-in conductors to insulators

5.1.0 Stringing Conductors

The **tension stringing method** is the method widely used for paying off a conductor from the supply reel. This method prevents conductors from touching the ground or sagging into nearby energized lines.

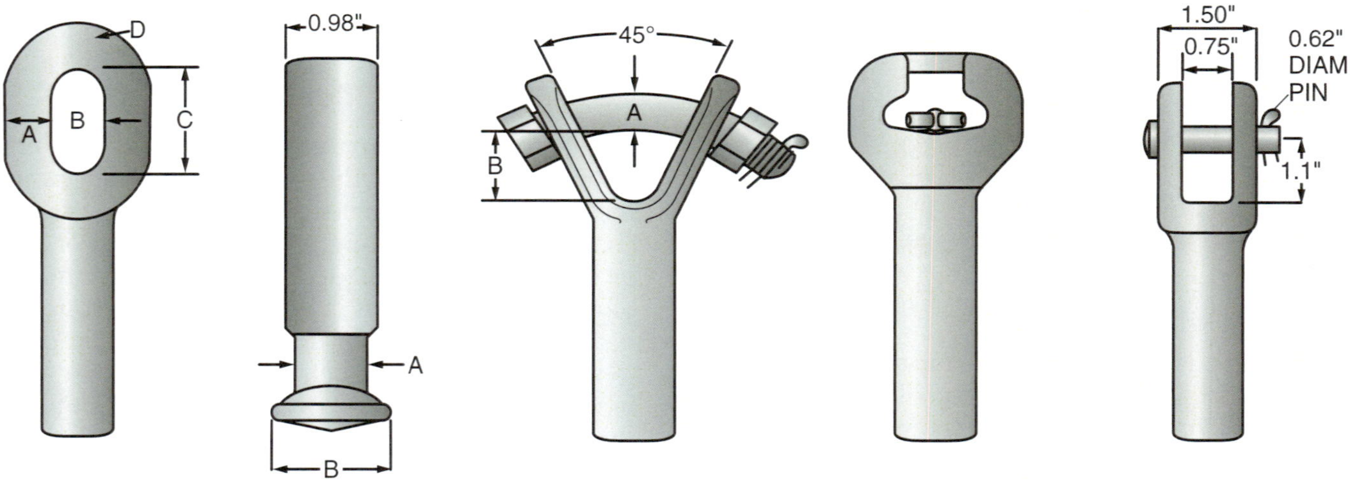

81203-11_F15.EPS

Figure 15 Insulator end fittings.

81203-11_F17.EPS

Figure 17 Stringing blocks.

A **bullwheel tensioner** (*Figure 18*) is located at the conductor supply reel. A **drum-style pulling machine** (*Figure 19*) is located at the end of the line. Both machines provide tension.

A wire mesh grip, called a **Kellems grip** (*Figure 20*) is used to attach the conductor to a pulling rope or cable. Do not use a bulldog grip or Chicago-style grip to attach the conductor to a pulling rope (bull rope). Both types can damage the conductor. A swivel is used to prevent twisting. The pilot line guides the pulling rope as it is passed through stringing blocks to the pulling machine. The rope, in turn, pulls the conductor off the reel and over the stringing blocks without touching the ground. When a reel of conductor runs out, a **double Kellems grip** is used to form a temporary splice between the end of the conductor on the empty reel and the conductor on the new reel. A temporary splice can pass through a stringing block. Not all permanent splices can pass through a stringing block. If each phase has multiple conductors, an **articulated running board** (*Figure 21*) is used to keep the conductors aligned as they are pulled through the stringing blocks. As the conductor is pulled, the pulling rope is wound onto a takeup reel. A shallow pulling angle must be maintained between the stringing block on the first transmission structure

On Site

Bundled Conductors

Each phase of a transmission voltage below 200kV is usually carried by a single conductor. Voltages above 200kV often use multiple (bundled) conductors for each phase. Two to four bundled conductors increase the current-carrying capacity of the line. They also reduce power loss.

81203-11_SA02.EPS

81203-11_F18.EPS

Figure 18 Bullwheel tensioner.

81203-11_F19.EPS

Figure 19 Drum-style pulling machine.

and the pulling machine. The slope must not be greater than a ratio of five horizontal to one vertical (*Figure 22*). Conductors being installed may pick up a static charge or an induced voltage from a nearby power line. They can also become accidentally energized. To protect against these voltages, always install a **running ground** (*Figure 23*) on any new conductor being installed.

WARNING!

Never stand beneath conductors that are being strung between transmission structures. Only trained and authorized persons can operate equipment used to string conductors.

NOTE

Wire mesh grips have several names. They include Kellems grip, basket grip, sock, and Chinese finger.

5.1.1 Stringing Shield Conductors

Shield conductors or wires, sometimes called ground wires, are installed on the raised peaks of transmission structures to protect transmission structures from lightning (*Figure 24*). Two ground wires are often used. These conductors are strung the same as other conductors on the transmission structure. One difference is that the stringing blocks are top-mounted instead of suspended. Today, many utilities are using an **optical ground wire (OPGW)** in place of a standard ground wire. This type of conductor has outer layers of steel and aluminum for strength and conductivity. The core of the conductor contains optical fibers (*Fig-*

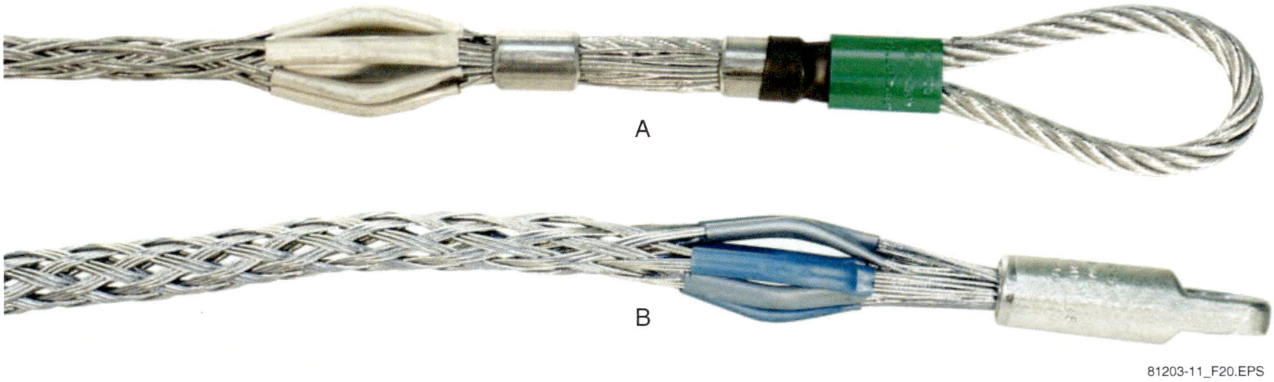

Figure 20 Kellems grip.

Figure 21 Articulated running board.

ure 25). The outer layers act as the ground, while the inner optical fibers are used for high-speed transmission of the utility's voice and data communications. Since the conductor contains optical fibers, different stringing blocks and other devices that are made for pulling this type of conductor may be required.

Figure 23 Running ground.

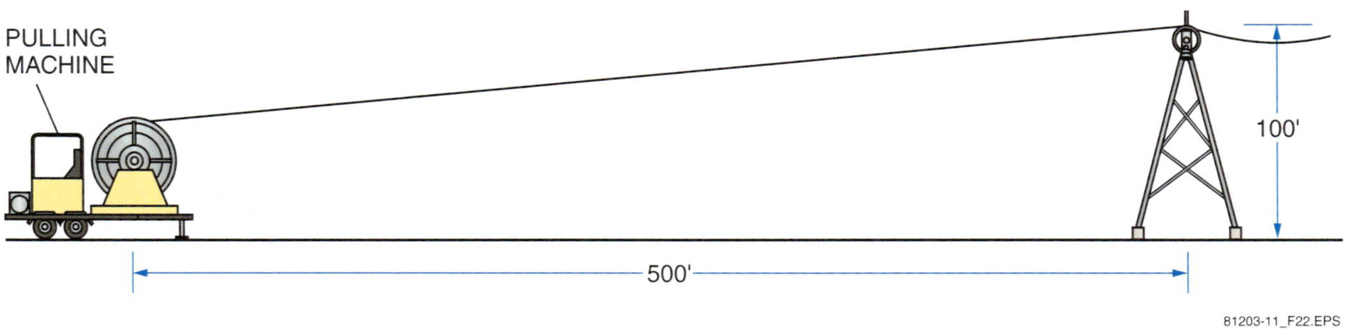

Figure 22 Pulling slope.

Figure 24 Shield wires.

> **NOTE**
> Machines are available that can wind a fiber optic cable around an installed conventional ground wire.

Extra care is needed when stringing OPGW so the optical fibers are not crushed. Travelers and blocks must have a protective lining. The minimum sheave diameter depends on the manufacturer's design. Always consult the manufacturer for minimum sheave and bullwheel diameters, pulling speeds, and pulling tension.

Many OPGW cables require the use of a swivel on the pulling line, followed by an anti-rotation device, in order to minimize the tendency of the OPGW to twist during stringing. Design improvements have produced some OPGW cables with installation tension requirements that may not need an anti-rotation device.

When replacing overhead ground wire (OHGW) with OPGW, the setup of the cable reel, tensioner, and pulley is critical. The slope of the cable should never be steeper than three-to-one

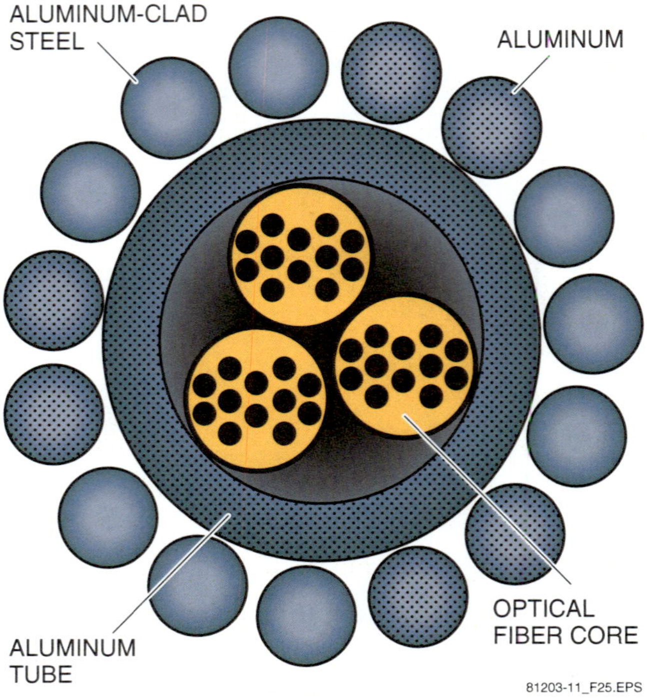

Figure 25 Optical ground wire construction.

(3 horizontal to 1 vertical). If terrain or obstructions prevent placing the tensioner in the proper location, the first sheave is lowered to obtain the three-to-one slope, and then raised during hardware installation.

A typical stringing tension is in the range of 15 percent or less of the OPGW rated breaking strength. The maximum stringing tension should be limited to 20 percent of the rated breaking strength. Pulling speeds should be in the range of 1.2 to 3.1 mph (2 to 5 km/h). Always consult the manufacturer's recommendations before stringing OPGW.

Establishing the sag in OPGW is much like the method used for conductors and overhead ground wires. Sag and tension information is supplied by the OPGW manufacturer. The greatest concern of line crews is not to crush the fiber optic strands. The grips must be specifically designed for OPGW. Manufacturers provide closely machined grips designed to match the diameter of the OPGW cable. With some OPGW cables, line crews can use formed wire grips if the manufacturer approves them.

6.0.0 Conductor Splicing

A fixed length of conductor is available on each reel. For that reason, long transmission lines contain splices, often called sleeves. During stringing, a double Kellems grip acts as a temporary

On Site

Transposition

The spacing and position of phase conductors in relation to other conductors on a transmission structure can cause a voltage imbalance at the end of the line. This problem can be corrected by transposing the lines to balance the system. A power line on which conductors are properly spaced and which is not too long probably does not need this to be done.

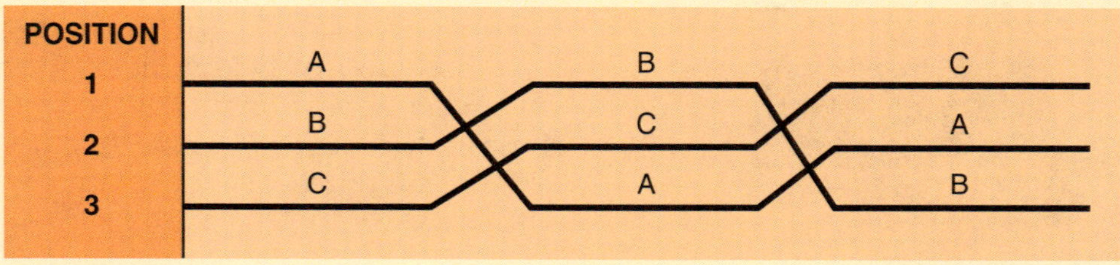

81203-11_SA03.EPS

splice. It allows the joined conductors to be pulled through stringing blocks. Some permanent splices are too stiff and their size makes it hard to pull them through a stringing block. Once the line is strung and the position of the splices are noted, permanent splices must be installed. Slack can be put in the line so that it can be dropped on or near the ground to make the splice. Splices can also be made in the air. In that case, the two conductors to be joined are held together by come-alongs that are attached to cable grips during the splicing. Workers typically perform the splicing from a bucket.

NOTE

Splices should be kept at least 50 feet away from a transmission structure and/or insulator. Any conductor lowered to the ground for splicing must be placed on a tarp or cribbing to prevent contact with the ground.

Conductors can be spliced in the air (mid-span) or on the ground. Splicing is commonly done with **compression sleeves**, which are **full-tension splices**. This means they have most of the strength and conductivity of the original conductor.

Compression sleeves are squeezed around the conductor with a hydraulic crimping tool (*Figure 26*). A splice for joining aluminum conductor steel reinforced (ACSR) conductor has two parts. An inner steel sleeve joins and strengthens the steel core. An outer aluminum sleeve joins the aluminum conductors and restores conductivity. To make the repair, the outer layer of aluminum is cut back evenly on each side of the splice to expose the core. The cut ends of the aluminum are taped to prevent unraveling of the conductors.

WARNING!

If a Kellems grip is used to hold the two ends of the conductor together, it is keeping the two ends at the same potential. Before removing the grip to make the splice, apply a jumper or other method between the two conductors to maintain continuity.

Before joining the core wires, slide the aluminum sleeve over one of the conductors. Crimp the steel sleeve to join the core wires. The conductors on both sides of the splice must be brushed to remove any oxide coating. After that, a corrosion inhibitor is applied to the cleaned aluminum before the outer sleeve is slid into a center position and crimped in place around the conductor. *Figure 27* shows a cutaway view of the completed splice.

NOTE

With some compression sleeves, corrosion inhibitor is injected into the outer sleeve.

7.0.0 Connectors

Different types of connectors are used when stringing a power line. Three common uses are to dead-end lines, make repairs, and to tap into an existing line. These connectors can be applied by compression, implosion, with a powder-actuated tool, or they can be bolted on.

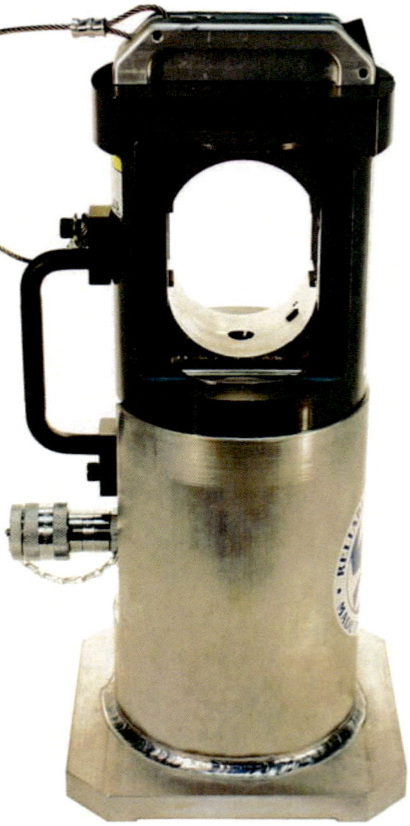

81203-11_F26.EPS

Figure 26 Hydraulic crimping tool.

7.1.0 Compression Connectors

In addition to standard compression terminals, there are various types of compression connectors used in transmission work. These include dead ends (*Figure 28*) as well as straight-through and tee-type sleeve connectors (*Figure 29*).

Compression connectors are applied by squeezing them around the conductor with a crimping tool. A manual or hydraulic crimping tool can be used to apply the connectors. These tools can be fitted with a variety of different jaws to handle different types of fittings. Hydraulic crimping tools are used more than manual tools because they can better exert the force needed to crimp the connector.

7.2.0 Powder-Actuated Tap Connectors

Connectors shaped like the letter C are used as tap connectors. Conductors are placed in the C-shaped body. Then a wedge is inserted to hold them in place. A powder-actuated AMPACT™ tool (*Figure 30*) is then used to drive the wedge into the connector body to firmly join the conductors (*Figure 31*). Dimples appear on the tap body and wedge to indicate an acceptable connection (*Figure 32*).

Follow these safety rules when using an AMPACT™ tool:

- Read the operator's manual before using the tool.
- Only trained and authorized persons can operate an AMPACT™ tool.
- Never operate an AMPACT™ tool that is damaged.
- Wear safety glasses and gloves when using a powder-actuated tool.
- Ensure that no one is in the line of fire when using the tool.
- Only use a hammer to strike the tool.
- Use the correct color-coded cartridge for the connector being installed.

8.0.0 Dead-End, Sag, and Clip-In Conductors

After phase conductors are strung between transmission structures, they must be dead-ended on one end, sagged, and dead-ended on the other end. After the line is dead-ended on both ends, the stringing blocks can be removed and the conductors clipped-in to the insulators.

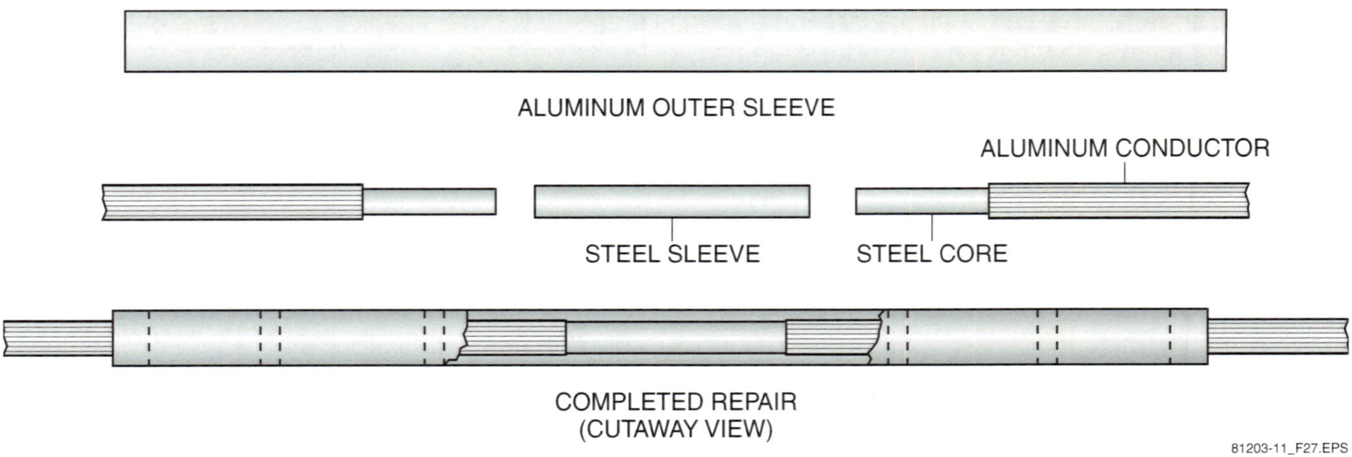

81203-11_F27.EPS

Figure 27 Cutaway of splice.

On Site

Implosive Sleeves

An implosive sleeve uses the pressure created during an explosion to compress a sleeve around a conductor to form a splice. The splice consists of an aluminum sleeve wrapped with an explosive. This photo shows a BURNDY® Implo® splice being prepared for use. Different sleeve types are used for different conductors. Splicing can be done on the ground or in the air. The conductor must be marked for proper insertion into the sleeve. Once the sleeve is prepared, the work crew is backed off at least 100 feet and the sleeve is ignited. Implosive sleeves use a non-electric detonator that is not affected by stray voltages. Its shape is such that the splice can be pulled through a stringing block with no damage to the splice or the stringing block. This eliminates the need to double-grip conductors when they are pulled. There are also compression terminals that can be applied using the same method.

81203-11_SA04.EPS

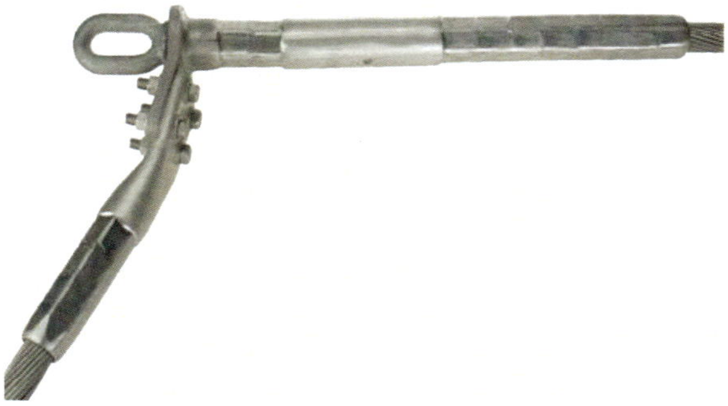

81203-11_F28.EPS

Figure 28 Dead-end compression connector.

8.1.0 Dead-End Conductors

Dead-ending is the process of securing the terminating ends of conductors at the transmission structures on each end of the section of new line. Once the line is strung between the transmission structures, it is dead-ended at the first transmission structure using a dead-end connector attached to a dead-end insulator (*Figure 33*). After the line is dead-ended at the first transmission structure, the length of conductor can be tensioned and sagged. Once that is done, the other end of the line can be dead-ended to the last transmission structure.

8.2.0 Sagging Conductors

Conductors are tensioned to achieve a certain amount of sag. The amount of sag in a line can be checked by sight using a scope, or it can be measured by timing how long it takes a wave on the conductor to be reflected back and forth between transmission structures.

8.2.1 Tension Sagging

The line to be tensioned and sagged is dead-ended at the first transmission structure with the conductor supported at each of the other transmission structures by stringing blocks (*Figure 34*). A pulling machine at the other end of the conductor pulls it to apply tension and bring the line up to the proper sag. Once the proper sag is achieved, the line is snubbed off to an anchor and left to set for at least two hours to creep.

On Site

Why Sag Conductors?

Lines are sagged to compensate for the effects of expansion and contraction of the conductor that take place as temperature rises and falls. If a line has too little sag, it could contract in winter to the point that it could stress components and cause them to fail. If a line has too much sag, it could expand and droop close to the ground in summer, where it would present a safety hazard.

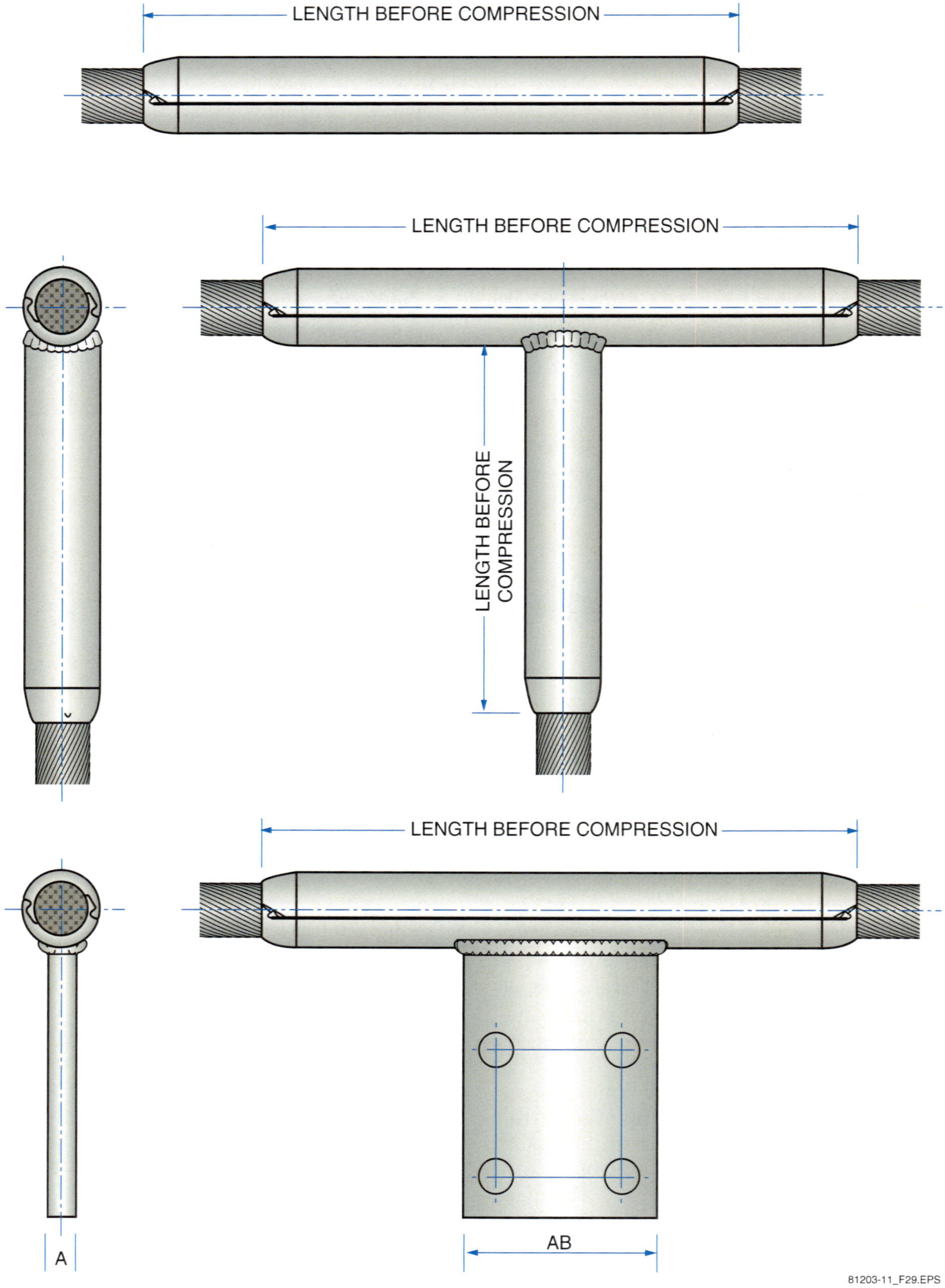

Figure 29 Sleeve-type compression connectors.

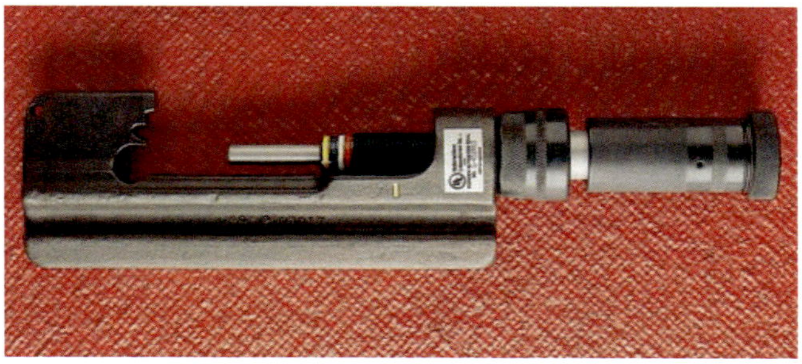

81203-11_F30.EPS

Figure 30 AMPACT™ tool.

During that time, tension equalizes and sag becomes evenly distributed between the transmission structures. The conductor stretches from its own weight and expands or contracts based on the ambient temperature.

A **dynamometer** (*Figure 35*) is used to measure pulling force when installed in the pull line. It is the preferred method used to measure tension as sag is adjusted. A sag chart (*Figure 36*) may be used to find the correct sag when tension, conductor size, span length, and temperature

81203-11_F31.EPS

Figure 31 Completed connection.

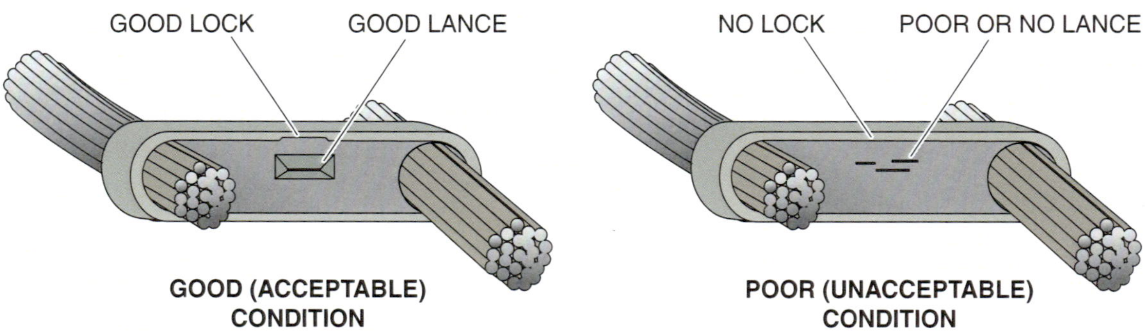

81203-11_F32.EPS

Figure 32 Good and poor indicators.

are known. To check the extent of sag, use a **sag scope** (*Figure 37*) or a surveyor's transit. A sag scope is similar to a telescopic sight on a rifle. It is attached to an end transmission structure and used to sight along the conductors. It can confirm that sag is correct or it can be used to make adjustments to sag. Creep in new conductors can be compensated by either over-tensioning or pre-stressing. With over-tensioning, the conductor is tensioned 5 to 10 percent over the required tension, using a dynamometer. Over time, the conductor creeps back down to the correct tension and sag height. Pulling the conductor slightly higher than the desired sag height can also over-tension the line. Pre-stressing greatly over-tensions the conductor for up to 30 minutes. Over-tensioning should never exceed 50 percent. A dynamometer is used to measure tension. After tension is released, the relaxed conductor is pulled to the correct sag height. Once sag is set, the minimum ground clearance for conductors must be maintained between all transmission structures.

81203-11_F33.EPS

Figure 33 Dead-ended conductor.

81203-11_F35.EPS

Figure 35 Dynamometer.

> **NOTE**
>
> An analog or digital dynamometer can be used to measure tension. A dynamometer is often built into the operator's console on the bullwheel tensioner. Always compensate for temperature when adjusting sag.

8.2.2 Sag Measurement by Timing

If a conductor that is strung between two transmission structures is struck, or pulled with a rope, it induces a wave on the conductor that travels back and forth between the two transmission structures. The time it takes for the wave to travel between transmission structures and the number of return waves counted, when entered into an equation, can be used to determine sag.

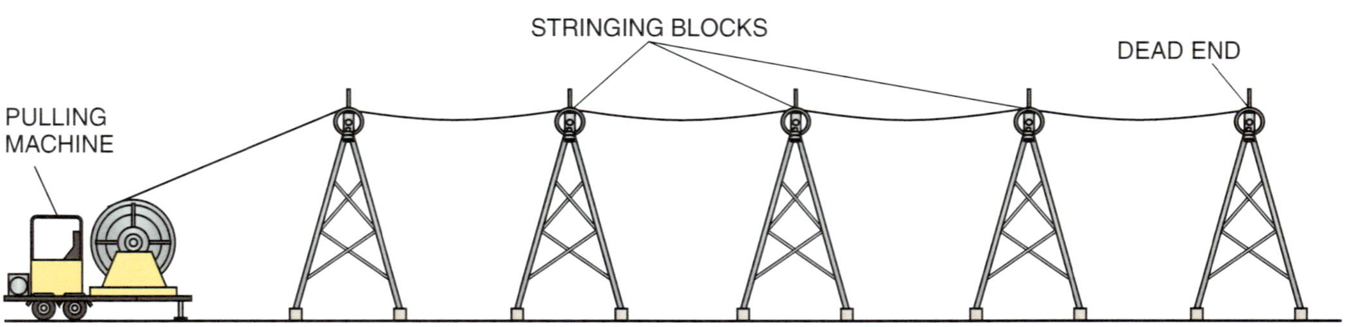

81203-11_F34.EPS

Figure 34 Tensioning the line.

RULING SPAN – 600 FEET				
	INITIAL		FINAL	
Conductor Temp (°F)	Tension (lbs)	Sag (ft)	Tension (lbs)	Sag (ft)
30 with 9lb/ft² wind	7,134	6.58	5,676	8.27
30	6,593*	4.90	4,789	6.74
35	6,434	5.02	4,630	6.97
40	6,275	5.14	4,479	7.21
45	6,117	5.28	4,334	7.45
50	5,960	5.42	4,196	7.70
55	5,804	5.56	4,064	7.95
60	5,650	5.71	3,940	8.20
65	5,498	5.87	3,821	8.45
70	5,347	6.04	3,709	8.71
75	5,200	6.21	3,603	8.97
80	5,055	6.39	3,503	9.22
85	4,913	6.57	3,408	9.48
90	4,774	6.76	3,317	9.74
95	4,639	6.96	3,232	10.00
200	2,731	11.84	2,380	13.60

81203-11_F36.EPS

Figure 36 Sag chart.

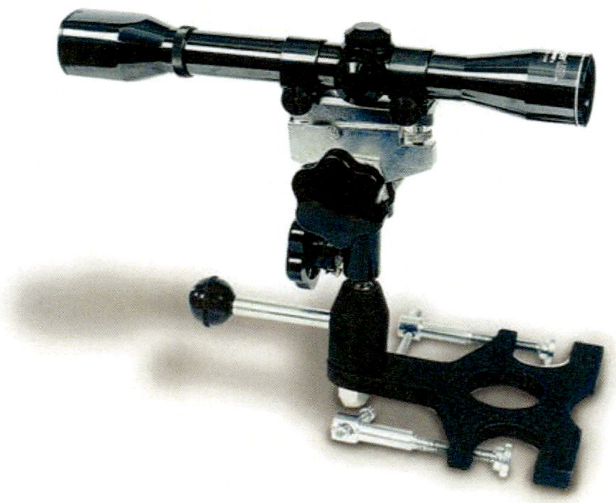

81203-11_F37.EPS

Figure 37 Sag scope.

$$\text{Sag D} = 48.3(t/2n)^2$$

The letter D is sag in inches. The letter t is the time it takes a wave to return in seconds. The letter n is the number of return waves counted. Assume it takes 12 seconds for a wave to return and four return waves are counted. In this example, the sag is 108.7 inches.

$$\text{Sag D} = 48.3(12/2 \times 4)^2$$
$$\text{Sag D} = 48.3(12/8)^2$$
$$\text{Sag D} = 48.3(1.5)^2$$
$$\text{Sag D} = 48.3(2.25)$$
$$\text{Sag D} = 48.3 \times 2.25 = 108.7''$$

A time-sag table (*Figure 38*) can also be used to determine sag. Timing should be done with a stopwatch.

NOTE

Use a rope to start a wave. Attach a rope from the conductor to the ground near the transmission structure. A sharp tug on the rope induces a wave. Do not attempt to check sag by timing on a windy day. The wind will move the conductor and affect the return waves.

8.3.0 Clipping-In Conductors

Conductors are attached (clipped-in) to the insulators after they have been installed, sagged, and dead-ended to transmission structures on both ends of the line. On a line that uses suspended insulators, this involves the following three steps:

- Removing the stringing block from the insulator or yoke plate
- Installing suspension clamp(s) on the insulator or yoke plate
- Securing conductor(s) to suspension clamp(s)

8.3.1 Removal of Stringing Blocks

Stringing blocks are under tension after the line has been sagged and dead-ended. That line tension must be relieved to remove the stringing

TIME SAG TABLE

Sag (Inches)	Return of Wave			
	3rd Time (Sec.)	5th Time (Sec.)	10th Time (Sec.)	15th Time (Sec.)
5	1.9	3.2	6.4	9.7
6	2.1	3.5	7.0	10.6
7	2.3	3.8	7.6	11.4
8	2.4	4.1	8.1	12.2
9	2.6	4.3	8.6	13.0
10	2.7	4.6	9.1	13.7
11	2.9	4.8	9.5	14.3
12	3.0	5.0	10.0	15.0
13	3.1	5.2	10.4	15.6
14	3.2	5.4	10.8	16.2
15	3.3	5.6	11.1	16.7
16	3.5	5.8	11.5	17.3
17	3.6	5.9	11.9	17.8
18	3.7	6.1	12.2	18.3
19	3.8	6.3	12.5	18.8
20	3.9	6.4	12.9	19.3
21	4.0	6.6	13.2	19.8
22	4.0	6.7	13.5	20.2
23	4.1	6.9	13.8	20.7
24	4.2	7.0	14.1	21.1
25	4.3	7.2	14.4	21.6
26	4.4	7.3	14.7	22.0
27	4.5	7.5	15.0	22.4
28	4.6	7.6	15.2	22.8
29	4.6	7.7	15.5	23.2
30	4.7	7.9	15.8	23.6
31	4.8	8.0	16.0	24.0
32	4.9	8.1	16.3	24.4
33	5.0	8.3	16.5	24.8
34	5.0	8.4	16.8	25.2
35	5.1	8.5	17.0	25.5
36	5.2	8.6	17.3	25.9
37	5.3	8.8	17.5	26.3
38	5.3	8.9	17.7	26.6
39	5.4	9.0	18.0	27.0
40	5.5	9.1	18.2	27.3
41	5.5	9.2	18.4	27.6
42	5.6	9.3	18.7	28.0
43	5.7	9.4	18.9	28.3
44	5.7	9.5	19.1	28.6
45	5.8	9.7	19.3	29.0
46	5.9	9.8	19.5	29.3
47	5.9	9.9	19.7	29.6
48	6.0	10.0	19.9	29.9
49	6.0	10.1	20.1	30.2
50	6.1	10.2	20.3	30.5
51	6.2	10.3	20.6	30.8
52	6.2	10.4	20.8	31.1
53	6.3	10.5	21.0	31.4
54	6.3	10.6	21.1	31.7

Sag (Inches)	Return of Wave			
	3rd Time (Sec.)	5th Time (Sec.)	10th Time (Sec.)	15th Time (Sec.)
55	6.4	10.7	21.3	32.0
56	6.5	10.8	21.5	32.3
57	6.5	10.9	21.7	32.6
58	6.6	11.0	21.9	32.9
59	6.6	11.1	22.1	33.2
60	6.7	11.1	22.3	33.4
61	6.7	11.2	22.5	33.7
62	6.8	11.3	22.7	34.0
63	6.9	11.4	22.8	34.3
64	6.9	11.5	23.0	34.5
65	7.0	11.6	23.2	34.8
66	7.0	11.7	23.4	35.1
67	7.1	11.8	23.6	35.3
68	7.1	11.9	23.7	35.6
69	7.2	12.0	23.9	35.9
70	7.2	12.0	24.1	36.1
71	7.3	12.1	24.2	36.4
72	7.3	12.2	24.4	36.6
73	7.4	12.3	24.6	36.9
74	7.4	12.4	24.8	37.1
75	7.5	12.5	24.9	37.4
76	7.5	12.5	25.1	37.6
77	7.6	12.6	25.3	37.9
78	7.6	12.7	25.4	38.1
79	7.7	12.8	25.6	38.4
80	7.7	12.9	25.7	38.6
81	7.8	13.0	25.9	38.9
82	7.8	13.0	26.1	39.1
83	7.9	13.1	26.2	39.3
84	7.9	13.2	26.4	39.6
85	8.0	13.3	26.5	39.8
86	8.0	13.3	26.7	40.0
87	8.1	13.4	26.8	40.3
88	8.1	13.5	27.0	40.5
89	8.1	13.6	27.1	40.7
90	8.2	13.7	27.3	41.0
91	8.2	13.7	27.5	41.2
92	8.3	13.8	27.6	41.4
93	8.3	13.9	27.8	41.6
94	8.4	14.0	27.9	41.9
95	8.4	14.0	28.0	42.1
96	8.5	14.1	28.2	42.3
97	8.5	14.2	28.3	42.5
98	8.5	14.2	28.5	42.7
99	8.6	14.3	28.6	43.0
100	8.6	14.4	28.8	43.2
101	8.7	14.5	28.9	43.4
102	8.7	14.5	29.1	43.6
103	8.8	14.6	29.2	43.8
104	8.8	14.7	29.3	44.0

81203-11_F38.EPS

Figure 38 Time-sag table.

On Site

Cable Distance Meter

The device shown here uses ultrasonic signals to measure cable sag, cable height distance, and clearance of overhead conductors.

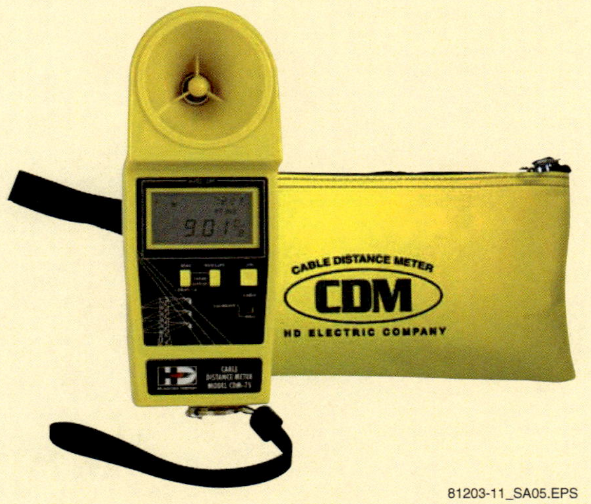

81203-11_SA05.EPS

block. Secure a come-along to the transmission structure above the conductor and near the stringing block. A strap is placed around the conductor and the come-along is used to lift and hold the conductor off the stringing block. With tension relieved, the stringing block can be removed from the insulator or yoke plate.

8.3.2 Installing a Suspension Clamp

A suspension clamp (*Figure 39*) is a cradle-type device on which the conductor hangs from the insulator or yoke plate. It may be made in pieces so that it can be taken apart for easier placement around the conductor. Before the conductor is clamped into the suspension clamp, it must be protected with **armor rods** (*Figure 40*), which are also called line guards. The armor rods are placed around the conductor where it passes through the suspension clamp and for several feet on either side of the clamp. It reduces flexing and chafing of the conductor at the clamp.

9.0.0 Completing the Conductor Installation

There are other steps that have to be taken once the conductors have been strung and clipped in. These other steps are designed to protect the installed conductors and ensure the safety of others

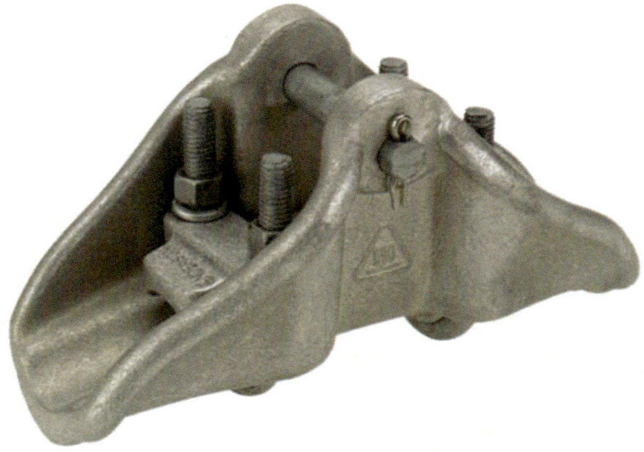

81203-11_F39.EPS

Figure 39 Suspension clamp.

who might venture near the installed power lines. These steps include the following:

- Installing dampers on conductors
- Installing spacers between bundled conductors
- Installing warning lights and day markers

9.1.0 Vibration Dampers

On long spans between transmission structures, wind can cause vibrations and/or oscillations to build up on a conductor. This unwanted movement can damage conductors. To reduce this movement, vibration dampers (*Figure 41*) are placed on the conductors. Their added weight helps suppress motion.

9.2.0 Spacers

Wind can also cause bundled conductors to contact each other. Spacers (*Figure 42*) are placed along the conductors to maintain the correct distance between them. The spacers are slipped over the conductors and tightened in place. Spacers for two-, three-, and four-conductor bundles are available. Spacers have to be placed along the bundled conductors. To reach the conductors, workers can use an aerial bucket. On higher transmission

GOING GREEN

Recycling

New power lines are often installed to replace outdated lines and transmission structures. When that is done, the conductors and the metal in the transmission structures can be recycled. The steel, copper, and aluminum that are recovered can be sold to provide the utility with a cash stream.

81203-11_F40.EPS

Figure 40 Armor rods.

81203-11_F41.EPS

Figure 41 Installed vibration damper.

structures, a **spacer cart** may be used. The cart is motorized and rides along the suspended conductors. The line worker installs the spacers from the safety of the cart. Note the large quantity of spacers hanging from the cart (*Figure 43*).

> **NOTE**
> Spacer dampers are available that function as both a spacer and a damper.

9.3.0 Warning Lights and Day Markers

The Federal Aviation Agency (FAA) requires tall structures, including some power transmission structures, to be equipped with warning lights to prevent aircraft from flying into them. The general rule is that anything over 200 feet requires a light, but shorter structures may be included, depending on location and other factors. The FAA also requires that power lines in areas where aircraft take off and land or where boats may pass under them be equipped with warning lights and/or day markers. Failure to maintain warning lights can result in a stiff fine. Self-powered warning lights (*Figure 44*) can be strung along a power line from a helicopter or a spacer cart. The lights are clamped around the conductor and receive their power by way of induction from the energized conductor.

81203-11_F42.EPS

Figure 42 Spacers installed.

81203-11_F43.EPS

Figure 43 Spacer cart.

81203-11_F44.EPS

Figure 44 Self-powered warning light.

To identify power lines in daylight, day markers (*Figure 45*) are installed along the line. Helicopters are used to install day markers in remote areas or where power lines cross water (*Figure 46*). Day markers are hollow spheres of different sizes. They are made of brightly colored lightweight plastic, and designed to quickly clip over the power line.

On Site

Monitoring Warning Lights

The FAA fines a utility if it allows transmission structure warning lights to go out. Automated monitoring with alarm notification is used to avoid this problem. This alerts the utility as soon as a warning light fails and allows them to dispatch a repair crew.

81203-11_F45.EPS

Figure 45 Day marker.

81203-11_F46.EPS

Figure 46 Installing day marker.

SUMMARY

Once a transmission structure is erected, workers have to install hardware and string conductors between the transmission structures. Installing power line conductors can be dangerous. Transmission structure heights often exceed 100 feet. Induced voltages can create a shock hazard. Conductors under tension can snap or break free if not properly secured. For those reasons, workers must maintain a safe work environment when performing these tasks.

Electrostatic and electromagnetic induction can induce a deadly electric current in a nearby conductor. A fault can energize grounded components. Workers can be protected from these hazards by working in an equal potential zone in which any surface they touch is at the same potential. Workers who are injured while working on a transmission structure must be removed in a timely and safe manner. Other workers may take part in some or all of the rescue. All workers must be trained to render simple first aid, perform CPR, and be able to aid in a transmission structure rescue.

In order for conductors to be strung between transmission structures, insulators are installed, along with stringing blocks. A pilot line is used to string the pulling rope through the stringing blocks. Conductors are strung between transmission structures using the tension stringing method. Conductors are spliced together in long pulls using compression sleeves. Different types of connectors are used on transmission lines to dead-end lines, make repairs, and to tap into an existing line. These connectors can be applied by compression or a powder-actuated tool. After conductors are strung between transmission structures, they must be dead-ended on one end, sagged, and dead-ended on the other end. Once that is done, the stringing blocks can be removed and the conductors clipped-in to the insulators. As a final step in conductor installation, spacers and dampers have to be installed on conductors, and warning lights and day markers have to be installed to prevent aircraft from flying into transmission structures or conductors.

Review Questions

1. Which of these devices operates by electromagnetic induction?
   a. Transformer
   b. Capacitor
   c. Resistor
   d. Conductor

2. The process by which an electromagnetic field from one conductor causes a current to be produced in a nearby object is known as _____.
   a. capacitance
   b. conduction
   c. induction
   d. osmosis

3. In an equal potential zone, the difference in potential between any two points in the zone is zero.
   a. True
   b. False

4. How is voltage affected by the distance from the point where an energized grounded conductor contacts the ground?
   a. Voltage is lowest at the point of contact.
   b. Voltage increases as distance increases.
   c. Voltage decreases as distance increases.
   d. Voltage is unaffected by distance.

5. If a worker on a transmission structure is injured and not breathing, the first thing you must do is _____.
   a. insert a breathing tube
   b. start mouth-to-mouth breathing
   c. lower the worker to the ground
   d. call 911

6. At what angle from vertical are two insulators installed to support one or more phase conductors?
   a. 30 degrees
   b. 45 degrees
   c. 60 degrees
   d. 90 degrees

7. The effects of corona on a power line can be reduced by installing a ____.
   a. corona ring
   b. capacitor
   c. transformer
   d. corona ground

8. A pilot line is used to _____.
   a. attach a conductor to a Kellems grip
   b. draw a bull rope through a stringing block
   c. attach a helicopter to a live phase conductor
   d. temporarily attach conductors to cross arms

9. Conductors are attached to a pulling line with a come-along.
   a. True
   b. False

10. The preferred method for installing conductors between transmission structures is _____.
    a. slack conductor stringing
    b. tension stringing
    c. stringing block pulling
    d. slack tension stringing

11. Shield conductors are installed on transmission structures _____.
    a. slightly below the conductors
    b. on the raised peaks of transmission structures
    c. perpendicular to the conductors
    d. along the bases of the transmission structures

12. How far away from a transmission structure or insulator must a conductor splice be kept?
    a. 200 feet
    b. 100 feet
    c. 75 feet
    d. 50 feet

13. A compression sleeve splice is applied around a conductor with a(n) _____.
    a. implosive clamp
    b. hydraulic crimper
    c. POT crimping tool
    d. impact wrench

14. The device circled in *Figure 1* is a(n) _____.
    a. vibration isolator
    b. H-connector
    c. tap connector
    d. running ground

15. If you have made a good connection when using an AMPACT™ tool, _____.
    a. the tool will make a loud click
    b. the connector will change color
    c. a dimple appears on the connector
    d. the connector will drop from the tool

16. Tension on a conductor is measured using a _____.
    a. sag scope
    b. surveyor's transit
    c. dynamometer
    d. strain gauge

17. The pre-stressing of conductors should never last more than _____.
    a. 6 hours
    b. 4 hours
    c. 1 hour
    d. ½ hour

18. The purpose of armor rods is to _____.
    a. protect insulators from vandalism
    b. limit flexing of conductors
    c. protect transmission structures from lightning
    d. prevent corrosion on conductors

19. Dampers are placed on power lines to reduce the damage that might be caused by _____.
    a. wind
    b. corona
    c. corrosion
    d. ice and snow

20. The FAA requires any structure to have a warning light if its height exceeds _____.
    a. 600 feet
    b. 400 feet
    c. 200 feet
    d. 150 feet

81203-11_RQ01.EPS

Figure 1

Trade Terms Introduced in This Module

Armor rods: Metal rods that are placed around a conductor as it passes through the suspension clamp. They are used to stiffen the conductor to prevent bending and flexing. They are often called line guards.

Articulated running board: A device used when pulling multiple (bundled) conductors through a stringing block. The running board allows all conductors to feed through the stringing blocks without becoming tangled.

Bullwheel tensioner: A machine that contains reels of conductor to be installed. As the conductor is unreeled, the bullwheel tensioner along with the pulling machine, maintains the correct pulling tension.

Compression sleeve: A type of conductor splice that is used to connect two conductors. It is squeezed around the conductors with a hydraulic crimping tool.

Corona ring: A doughnut-shaped device that is often installed on insulators on high-voltage lines to reduce the effects of corona.

Double Kellems grip: A double-ended wire mesh device that is used to make a temporary splice between conductors that are being pulled through stringing blocks. A double Kellems grip is used because some permanent splices cannot be pulled through stringing blocks.

Drum-style pulling machine: A powered pulling machine on which the pulling rope is wound as it pulls a conductor through stringing blocks. It is used with a bullwheel tensioner to help maintain proper line tension during stringing.

Dynamometer: A device used to measure tension on conductors as they are being pulled through stringing blocks.

Electromagnetic induction: It is the principle that allows a transformer to function. When the fluctuating electromagnetic field around a conductor that carries alternating current cuts through a nearby conductor, an alternating current is induced in the other conductor.

Electrostatic induction: It is the principle that allows a charged object to induce a static charge in a nearby object. When an object containing a charge is brought next to an object with no charge, it draws ions with the opposite charge to the surface and creates a static charge between the two objects.

Equal potential zone: A work area in which the potential (voltage) on all surfaces that can be reached or touched by a worker are equal. The zone is established by having all surfaces grounded at a common point with temporary ground jumpers.

Full-tension splice: A type of splice meant to restore strength and conductivity to a conductor. Compression splice sleeves are widely used.

Ground mat: A conductive mat placed on the ground to establish an equal potential zone. The mat is typically used to eliminate touch and step potential hazards.

Kellems grip: A wire mesh grip used to attach a pulling rope to the conductor so that it can be pulled through stringing blocks. Other names for a Kellems grip include basket grip, sock, and Chinese finger.

Minimum approach distance: A working distance to energized objects. Outside the minimum approach distance, no rubber gloves or sleeves are required, and conducting materials and tools can be handled. At distances less than the minimum, rubber gloves and sleeves must be worn and protective shields and nonconducting tools must be used. The minimum approach distance varies. As voltage increases, so does distance.

Optical ground wire (OPGW): A type of shield wire used to protect transmission structures from lightning. It has outer layers of steel and aluminum for strength and conductivity and an inner core that contains optical fibers. The outer layers act as the ground, while the inner optical fibers are used for the utility's voice and data signals.

Pilot line: A lightweight line used to draw a pulling rope through stringing blocks.

Rescue stretcher: A stretcher designed to lower badly injured and/or unconscious workers from high places.

Running ground: A temporary ground jumper equipped with rollers that keeps the conductor grounded while it is being pulled through stringing blocks.

Sag scope: An optical device similar to a telescopic sight on a rifle. It is mounted on a transmission structure and used to sight along power lines to check for correct sag.

Spacer cart: A lightweight cart that can travel between transmission structures while suspended from bundled conductors. Workers riding in the cart can install spacers or do other tasks.

Step potential: The potential (voltage) that exists between a person's feet when that person stands near an energized grounded conductor such as a transmission structure leg. Eliminate step potential by setting up an equal potential zone.

Stringing block: A pulley attached to insulators through which conductors are pulled while being installed. They are often called travelers.

Suspension clamp: A clamp hung below the insulator or yoke plate that supports and grips the conductor after it has been sagged and dead-ended.

Tension stringing method: A method used to string conductors between transmission structures. The conductor is pulled under tension over stringing blocks on each transmission structure. The method keeps conductors off the ground to avoid damage. It is the preferred method for installing new conductors.

Touch potential: The potential (voltage) that exists between a worker and an energized grounded conductor such as a tower leg. Eliminate touch potential by setting up an equal potential zone.

Yoke plate: A plate mounted directly below a suspended insulator(s) on which one or more suspension clamps can be attached.

Additional Resources

This module presents thorough resources for task training. The following resource material is suggested for further study.

Electric Power Transmission Systems, 2nd Edition, Upper Saddle River, NJ: Prentice Hall.

Figure Credits

Terry Williams, Module opener, Figures 2, 14 (bottom photo), 17, 18, 42, 45, 46, and SA02

A.B. Chance Co./Hubbell Power Systems, Figure 8

Yates Gear, Inc., Figure 10

Hubbell Power Systems, Figure 11, 15, 28, and 39

Topaz Publications, Inc., Figures 12, 14 (top photo), 16, 24, 33, and 41

James Anthony, Figure 13

Sherman & Reilly, Inc., Figures 19 and 21

Courtesy of SAFETYTEST, Figures 20, 23, and 26

Burndy LLC, SA04

AFL, Figure 29

Tyco Electronics Energy Division, Figures 30–32

Dillon Force Measurement, Figure 35

Condux Tesmec, Figure 37

USDA Rural Development, Figure 38

KT Power, Inc., Figure 40

Suparule Systems Limited, SA05

Morpac Industries, Inc., Figure 43

Calzavara SpA – Clampco Sistemi, Figure 44

NCCER CURRICULA — USER UPDATE

NCCER makes every effort to keep its textbooks up-to-date and free of technical errors. We appreciate your help in this process. If you find an error, a typographical mistake, or an inaccuracy in NCCER's curricula, please fill out this form (or a photocopy), or complete the online form at **www.nccer.org/olf**. Be sure to include the exact module ID number, page number, a detailed description, and your recommended correction. Your input will be brought to the attention of the Authoring Team. Thank you for your assistance.

Instructors – If you have an idea for improving this textbook, or have found that additional materials were necessary to teach this module effectively, please let us know so that we may present your suggestions to the Authoring Team.

NCCER Product Development and Revision

13614 Progress Blvd., Alachua, FL 32615

Email: curriculum@nccer.org

Online: www.nccer.org/olf

❑ Trainee Guide ❑ AIG ❑ Exam ❑ PowerPoints Other ______________________________

Craft / Level: Copyright Date:

Module ID Number / Title:

Section Number(s):

Description:

Recommended Correction:

Your Name:

Address:

Email: Phone:

81204-11

Transmission System Maintenance

Module Five

Trainees with successful module completions may be eligible for credentialing through NCCER's National Registry. To learn more, go to **www.nccer.org** or contact us at **1.888.622.3720**. Our website has information on the latest product releases and training, as well as online versions of our *Cornerstone* newsletter and Pearson's product catalog.

Your feedback is welcome. You may email your comments to **curriculum@nccer.org,** send general comments and inquiries to **info@nccer.org**, or fill in the User Update form at the back of this module.

V.1 12/11

81204-11
Transmission System Maintenance

Objectives

When you have completed this module, you will be able to do the following:

1. State the safety guidelines for working on transmission lines.
2. State the requirements for performing inspection and maintenance on insulators.
3. State the requirements for performing inspection and maintenance on conductors and conductor hardware, including switching, tagging, and clearance procedures.
4. State the requirements for performing inspection and maintenance on transmission structures and guys.
5. Explain the environmental issues that affect transmission line maintenance.
6. Prepare reports to document inspection results.
7. Perform inspection and maintenance on poles, transmission structures, and guys.
8. Perform fault location on transmission lines.

Performance Tasks

Under the supervision of your instructor, you should be able to do the following:

1. Prepare reports to document inspection results.
2. Perform inspection and maintenance on poles, transmission structures, and guys.
3. Perform fault location on transmission lines.

Trade Terms

Clearance
Compliance Management Plan (CMP)
Creepage
Creepage distance
Cycloaliphatic
Dry band
Flashover
Herbicide
Hydrophilic
Hydrophobic
Isolation guarantee
Landing signal operator (LSO)
Overhead ground wires (OHGW)
Radio frequency identification (RFID) sensor
Supervisory Control and Data Acquisition (SCADA)
Spalling
Switching order

Industry Recognized Credentials

If you're training through an NCCER-accredited sponsor you may be eligible for credentials from NCCER's Registry. The ID number for this module is 81204-11. Note that this module may have been used in other NCCER curricula and may apply to other level completions. Contact NCCER's Registry at 888.622.3720 or go to www.nccer.org for more information.

Contents

Topics to be presented in this module include:

Figures and Tables

1.0.0 Introduction

Transmission lines travel many miles across the country on long expanses of cleared land, connecting generating plants, switchyards, and substations. At switchyards, circuits are connected to the power grid. The grid is designed to route the electricity via several paths to the distribution substations. Providing a number of routes to the same substation ensures that a supply of electricity is always available to customers. It also allows utilities to isolate transmission lines for maintenance without disrupting power to customers.

Transmission lines have very few components. They consist of the transmission structure, conductors, insulators, and **overhead ground wires (OHGW)**. Because of their length, all transmission lines have splices on the conductors. Depending on the conditions, vibration dampers, spacers, and corona rings may be part of a transmission line. Insulators are key components on transmission lines. The size and length of the insulators depends on the voltage being carried by the line.

A large portion of the electrical transmission and distribution system in the United States is more than 40 years old. Many of the components have a lifespan of 40 to 50 years. Power outages are increasing because of a rise of equipment failures. Rising maintenance costs, increased usage, and insufficient capacity, coupled with a desire to minimize maintenance outages, have led many utility companies to operate under a "run it until it breaks" maintenance policy.

1.1.0 Inspection and Maintenance

Workers who inspect and maintain transmission lines are generally part of a work crew like the one shown in *Figure 1*. Major maintenance projects are often done in remote areas and may involve many workers, trucks, helicopters, and other equipment. Transmission line maintenance projects may require days and even weeks to complete. Careful planning is essential for a successful project. Job planning engineers need detailed inspection reports that accurately describe problems and identify the replacement parts needed to do the work. Job planners may have to arrange for the staging of equipment and parts in advance of the maintenance. Supervisors have to juggle people, equipment, and materials in order to schedule work crews. Managers constantly reassess budgets to ensure that enough people, equipment, and supplies are on hand to cover both planned and corrective maintenance. All of these factors, plus the need to provide continuous power, figure in any maintenance effort.

81204-11_F01.EPS

Figure 1 Work crew performing maintenance on a 500kV transmission line.

2.0.0 Safety

This section covers some key items of particular concern to transmission structure maintenance crews. *OSHA 1910.269 Appendix B, Working on Exposed Energized Parts* offers guidelines that explain safe work procedures. Here are some additional safety guidelines:

- A fall from a transmission line structure is almost always fatal. Many factors make climbing transmission structures dangerous, including slippery shoes and slippery steel, difficult climbs, and bad weather. Using properly inspected fall arrest equipment is the only way to ensure protection from falling.
- Never climb or crawl across an insulator. They are unstable and may be slippery from bird droppings and pollution. Serious cuts and lacerations can occur from a broken insulator. Use a ladder or platform for work around an insulator string.
- Do not stand under a transmission structure while work is going on above. Falling objects can penetrate hard hats or strike some other part of the body.
- Use a tag line to prevent material being sent up a transmission structure from spinning or striking the transmission structure supports. Do not remove a tag line until the load is securely in place. Workers on the tag line should position themselves away from the path of any falling items.
- Use safety lanyards on tools so they do not fall. If a tool starts to fall, do not try to reach for it; let it go. Crew members assisting on the ground can send it back up.

2.1.0 Insulator Testing

The insulation strength of insulator strings with broken insulators is unknown, but it is usually less than their design strength. Consider all damaged insulators in an insulator string to have no insulating value. The presence of live-line tools and ladders may provide an arcing path for current flow, further reducing the insulating strength of a string with broken insulators.

When working near insulators, line crews should test them for breakdown. There are battery-powered portable testers available that permit quick, safe, and reliable testing of energized insulators (*Figure 2*).

A line worker tests the insulator by moving the tester across the insulator string. A buzzer and an indicator light signal a warning when the tester locates a bad insulator.

2.2.0 Voltage Surges

Dangerous surges form when there is an electrical overstress on transmission line components. Surges are transient voltages that last for only a few milliseconds. Lightning is the most common cause, but the normal switching operation of capacitors can also create a surge. Other causes are faulty equipment placing a short circuit on the line and surges from the discharge of a lightning arrestor far away from the work site. *OSHA 1910.269 Appendix B, Working on Exposed Energized Parts* offers guidelines that can alert crews to safe work.

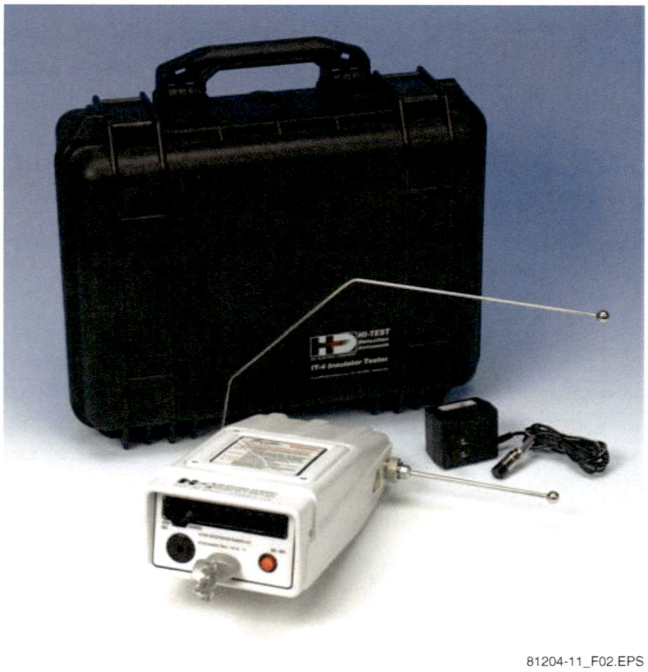

81204-11_F02.EPS

Figure 2 Portable insulator tester.

- Among the steps line crews can take to ensure their safety is a temporary reduction of surge gap distance within a half mile of the work site. Installing a protective surge gap on an adjacent transmission structure is acceptable. Lightning striking up to six miles away can cause a voltage surge that exceeds the capacity of insulation. A **flashover** is still possible; however, the flashover occurs across the protective gap, not between the worker and the line.
- Prevent automatic reclosing on the line where energized work is under way. There are two safety reasons for doing so: it prevents a surge caused by a worker mistake; and it protects against a transient overvoltage from switching on the line. If system-operating requirements prevent blocking of automatic reclosing, use some other method consistent with system design to control switching surges.
- Beware of surges on an adjacent line. A surge there can cause a dangerous overvoltage condition on the line in work. Always take into account the coupling from an adjacent line when figuring the minimum approach distances.
- If a line is to be isolated from the power source, make sure the proper **clearance** procedure has been implemented.
- Static electricity can jolt a climber and cause a fall when transitioning from an insulated ladder onto steel. Static shocks are also possible if there is inadvertent contact with ground while climbing a wood transmission structure. The only sure protection is a fall arrest system. Be observant and prepared for static shock. A fast slap of the hand against a grounded object will reduce the effects of a shock, as well as the risk of inadvertent movement.

2.3.0 Helicopter Safety

Helicopters have become an essential tool for transmission line maintenance. They provide speed and mobility no matter what the terrain or environment, overcoming access difficulties to remote areas. They can transport work crews and equipment more quickly and efficiently. In addition, maintenance work can often be done more quickly because workers can access the lines from the helicopter. Helicopters provide equipment spotting and serve as aerial support for ground workers. Although they are expensive to operate, in most cases the advantages of helicopters greatly outweigh the costs.

Power companies use helicopters for inspections and patrols of transmission lines (*Figure 3*). Among the inspection services possible with he-

81204-11_F03.EPS

Figure 3 Transmission line inspection with a helicopter.

81204-11_F04.EPS

Figure 4 Transmission line maintenance from a helicopter.

licopters are detailed visual inspections, standard aerial patrols, infrared inspections, hands-on inspections, and transmission structure inspections. A knowledgeable worker with a pair of binoculars can inspect many miles of transmission line in a single day from a helicopter. Helicopters have proven invaluable for many maintenance functions (*Figure 4*).

Helicopters can be used for reconductoring, pole setting, transmission structure erection, insulator changeout, repairs, and heavy lift operations. Heavy lift services include transporting sections of transmission structures, poles, cross arms, and insulator strings.

2.3.1 Helicopter Operations

Operating with helicopters, though routine, is dangerous. When flying next to power lines, factors such as winds, workers outside the aircraft, external loads, and weather, can create potential hazards. Thorough work planning that clearly outlines job tasks in step-by-step detail listing individual responsibilities, emergency procedures, and a weather briefing, ensures helicopter work remains safe. Refer to *FAA Regulations Part 133* and *Advisory Circular AC 133-1A, Rotorcraft External-Load Operations* for detailed regulations for helicopter operations.

2.3.2 Helicopter-Specific PPE

Line workers wear PPE when working with helicopters, as with any other maintenance task, but rotor downwash adds requirements. Eye protection is necessary to protect from flying debris. Hard hats must have a chinstrap. Many helicopter-borne workers wear an aviator cranial helmet with chinstrap fastened, goggles in place, and communication with the pilot. Workers around helicopters must wear tight-fitting clothes; flapping clothing can get caught in the hoist line or snag on uneven, jagged transmission structure surfaces. A fall-arrest system must be used when standing on the helicopter skid or working from a platform outside the helicopter. Tools should be on lanyards to prevent them from falling and injuring someone on the ground. If voice communication between the workers and the pilot is not available, hand signals are used to ensure effective communication with the pilot.

Line workers often stand or sit on the landing skids or a work platform while performing maintenance (*Figure 5*). Working from the side of a helicopter places the workers in the downwash from the rotors and they are subjected to loud noise and cold. Dressing warmly and using good hearing protection is essential. In cold weather, workers may need to wear exposure suits.

81204-11_F05.EPS

Figure 5 Workers installing armor rod from a helicopter.

2.3.3 Helicopters on the Ground

Workers must approach helicopters from the side or the front, but always within the range of the pilot's vision. Spinning rotors can sag at low speed, so they are not always the same height above the ground. Walk crouched over when under the rotor, even when it appears well above the head. Stay far away from the tail rotor. If suddenly blinded by dirt, sit down and wait for assistance. Never try to grope or feel the way to or from the helicopter.

If the takeoff site is on a hill, never approach or depart the helicopter on the upslope side. Also, never enter the area of lowest rotor clearance. To avoid contact with the main rotor blade, carry long pieces of equipment or tools at or below the waist. Never carry long equipment or tools upright or over the shoulder.

The downwash area can be as much as 100 feet from a helicopter on the ground. Do not set any loose equipment in the downwash area. A seemingly harmless object can become a dangerous unguided missile, particularly from the strong downwash during a takeoff. Helicopter engines are subject to foreign object damage from a small object being sucked into the engine intake.

2.3.4 Helicopter Hoisting Operations

Helicopter hoisting operations are inherently dangerous. All ground personnel must be trained in the proper procedures. They should know the hand signals used during an external load operation. There should be a **landing signal operator (LSO)** in a position visible to the pilot and the hookup person. The LSO stands in front of the aircraft and to the right of the flight path and as close to the operating area as possible with his/her back to the wind. *Figure 6* is a chart of the FAA-approved standard helicopter hoisting operation hand signals.

> **WARNING!**
>
> Do not hook or unhook a load unless properly trained. High levels of static electricity can build up in loads suspended beneath helicopters. Never touch an ungrounded suspended load. Ground loads by touching the hoist cable using a grounded insulated tool with a pull-away clamp. If possible, the LSO will direct the pilot to set the load on the ground in order to discharge static electricity before anyone touches it.

After hooking up the cargo sling, ground personnel must move clear of the aircraft. This prevents load cables from entangling a ground crew member and hoisting him or her aloft. Staying well clear also ensures that a swinging load does not strike a ground worker. After workers are clear, the LSO can signal the pilot to take off.

In some cases, workers are in a basket suspended well below the helicopter. All baskets and hoist cables must be inspected in accordance with *FAA Part 133*.

2.3.5 Fueling

Fueling a helicopter in the field becomes a routine operation when workers know and follow safety precautions. All components of the refueling process must be grounded to the same potential to prevent sparks from static electricity.

- Establish a good ground by driving a conductive rod into the ground.
- Ground the fuel tank and fuel pump to the ground rod.
- Ground the helicopter to the ground rod.
- Touch the nozzle to the helicopter before inserting it into the fuel inlet. Approved fuel hoses have an internal ground bonded to the fuel pump.

Stop refueling immediately if there is lightning in the area. Also, do not smoke within 100 feet of refueling operations.

2.3.6 Pilot Responsibility

Operating limitations, weather, and all aspects of flight safety are always the responsibility of the pilot in command of the helicopter. Typically, if the cloud ceiling is at least 800 feet with visibil-

Figure 6 Helicopter hand signals.

ity of 2 miles (3 miles in mountains), it is legal to fly. Some helicopter companies and utilities have more stringent safety policies than the FAA minimums. The pilot has the final say and may at times cancel a flight that is above FAA or company minimums.

2.3.7 Helicopter Grounding

The ungrounded helicopter and all wash apparatus components are electrically bonded together. This permits the water spray used to wash insulators from the helicopter to have prolonged safe contact with energized high voltage transmission lines. An ungrounded, fully bonded system does not have any adverse effect on the transmission line system, the aircraft, or operators. To ensure that the airborne washing operation is safe, the work crews must follow the safety policies of the utility company and helicopter operator, as well as FAA regulations.

3.0.0 Insulator Inspection and Maintenance

Insulators serve two purposes. They attach the conductor to the transmission structure and they insulate power lines from ground. *Figure 7* shows insulators on a steel tower. The groups of four insulators are strain insulators. They hold the weight of the conductor span. The single insulator mounted vertically is a suspension insulator. It holds the weight of the conductor that passes under the transmission structure to the next span.

There are two major design criteria of which inspectors must be aware. First, insulators must be mechanically strong to counter the strain of a long conductor span. Second, insulators must have an efficient electrical design in order to counter conditions such as **creepage**, flashover, and arcing.

A successful insulator inspection depends on the trained eye of the inspector. Experienced inspectors can quickly recognize damage. Most insulator damage is caused by high winds, lightning flashovers, contamination flashovers, and vandalism. A weakened insulator may not be strong enough to hold up conductors, particularly on long, heavy spans over mountains, at dead-end line transmission structures, and at sharp bends in a transmission line. It is important to replace damaged insulators as soon as possible after they are identified.

81204-11_F07.EPS

Figure 7 Transmission structure insulators.

3.1.0 Inspection Guidelines

Glass and porcelain insulators can be punctured by severe electrical stress from flashover and current leakage. A manufacturing process of thermal cooling ensures that the glass shed shatters if punctured. The mechanical integrity of the insulator is unaffected, but a damaged disc is clearly visible (*Figure 8*). Even though the insulator still works, line workers have to replace them because there is a shorter leakage length between the metal ends of the insulator. In the event of contamination, there is less resistance, making flashover more probable.

Inspect glass insulators for shattering. Prolonged arcing erodes the surface layer of glass insulators. That makes the insulator susceptible to arcing and corona. The insulator becomes brittle and eventually fractures. Gunshot damage has the same effect.

Look for insulator puncture damage on porcelain pin-type insulators. Puncture damage exposes the interior rod to moisture, which can lead to brittleness and fracture. Do not forget to look on the ground beneath the insulator for pieces of glass and porcelain.

When moisture penetrates the bonding material of some types of insulators the bonding material undergoes a chemical change. The moisture expands in freezing temperatures. The pressure from the expansion works its way through minute internal cracks and may cause the insulator to explode.

Look for evidence of excessive arcing, which leaves tracks of carbon on the surface of the insulator. The carbon tracks are highly conductive and provide a low-resistance path for flashover.

81204-11_F08.EPS

Figure 8 Damaged insulator bell.

Inspect fittings on the end of insulators for corrosion that destroys the metal. If not repaired, conductors may fall to the ground or insulators may fall off the transmission structure. Products of corrosion that drip onto the insulator sheds react chemically and further deteriorate the insulator.

3.2.0 Composite Insulators

Cycloaliphatic resin is a lightweight epoxy material surrounding a fiberglass center rod with a silicone rubber coating. As these insulators age in severe environments, sharp points on the rough surface can ionize the surrounding air. The result is a decrease of the electrical strength and service life of the insulator. Inspect **cycloaliphatic** resin insulators for rough surfaces that are the effects of aging. Additional silicone rubber can be spread on the surface to smooth out the rough points. If that is not possible, the insulator is replaced.

Problems may occur when cycloaliphatic resin insulators are used on very high-voltage lines (200kV and above) and are under heavy mechanical loads. If damaged, rain and pollutants gain access to the surface of the fiberglass center rod. Electrical discharges in the presence of water create acids that weaken the center rod. Erosion, melting, burning, and displacement of both the fiberglass rod and the polymer sheds can occur.

When the center rod is under mechanical stress in conjunction with the weakening, brittle failure occurs. Brittle failure mode is comparable to the process of stress corrosion. The combined action of electrical activity and the presence of minor amounts of pollutants are able to degrade the surface of glass fibers and produce stress corrosion. The corrosion gives the appearance of acid damage. Mechanical stress, even though it may be much less than the breaking point stress, will snap the insulator in half.

3.3.0 Insulator Testing

A visual inspection will not find all insulator faults. A faulty insulator can be the source of intermittent outages that are very difficult to troubleshoot. Battery-powered portable testers enable line workers to test energized insulators in order to find hidden damage. A typical tester outputs a very large potential, as much as 10kVDC, on the insulator sheds when they are between the two probes. There is no shock hazard because a current limiter circuit protects the user from the large potential. *Figure 9* shows a tester that fits on a hotstick.

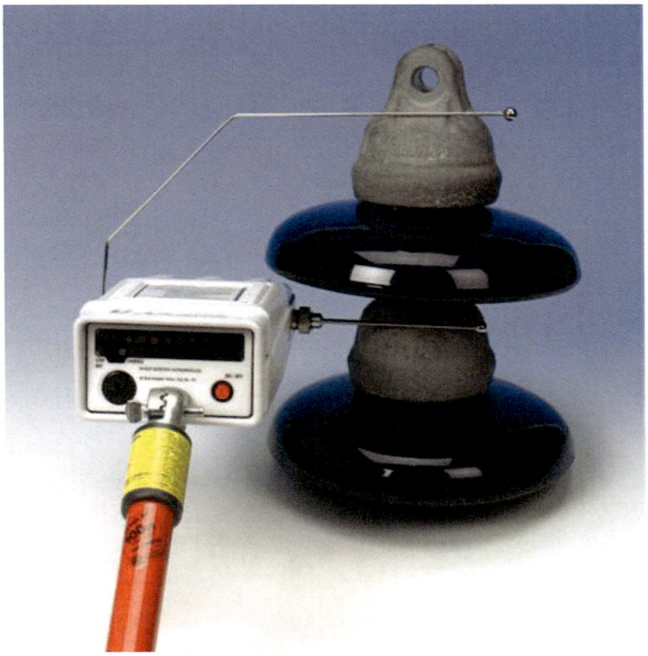

81204-11_F09.EPS

Figure 9 A typical portable insulator tester.

A line worker tests the insulator by moving the tester across the insulator string. The tester probes are positioned around each shed to permit the testing potential to pass through the insulator body. A buzzer and an indicator light signal a warning when the tester locates a bad insulator.

3.4.0 Corona Discharge

Damaged insulators can cause corona discharge. Corona discharge results when the broken part of the insulator has a pointed shape. Corona is plasma and usually has a purple glow, a hissing sound, and an odor. Corona causes power loss, noise, and radio frequency interference (RFI). Bolts and other pointed objects can also cause corona. A highly concentrated field forms at the tip of the point. The high potential from the concentrated field ionizes the surrounding air. If allowed to continue, corona weakens the insulator further. *Figure 10* shows line workers installing corona shields over bolt ends.

The transmission line carries alternating current. The expanding and collapsing magnetic field around the conductor affects the plasma and causes the air to vibrate. The vibration creates the hissing or crackling sound that is often heard around transmission lines. If the insulator is not replaced, the conditions around the sharp point might allow the area of ionization to continue growing and form a completely conductive path. This results in flashover.

81204-11_F10.EPS

Figure 10 Installing 230kV corona shields bare hand over splice clamp keeper bolts.

3.5.0 Insulator Contamination

Many transmission lines are located in areas of high pollution. Salt, dust, sand, chemicals, cement dust, smoke particles, and other pollutants collect on insulator surfaces. Contamination can also occur on the bottom sides of insulator bells. Rain may wash the contamination from the top, but cannot clean the underside.

3.5.1 Effects of Contamination and Pollution

Deposits on insulators generally take two forms, salt or pollution. In coastal areas, insulators receive a covering of sea salt and wind-driven contaminants heavy with salt. The salt layers become conductive in high humidity and fog. Pollution generally is in the form of smoke and other stack emissions, along with windborne dust from industrial plants. These pollutants are dry when deposited, but become conductive when wet. Industrial pollutants also deposit corrosive acid on insulators.

When the contaminants are wet, the surface of the insulator has a lower resistance, which can significantly reduce the effectiveness of the insulators. Creepage occurs on the surface area between the metal ends of the insulator. When creepage exists, the surface resistance can decrease enough that leakage currents flow along the outside of the insulator. Damage and contaminants from pollution can electrically shorten **creepage distance**.

Contamination on insulators causes the insulator surface to become **hydrophilic** or saturated in moist conditions. Combined with limited creepage distances, the moisture causes leakage currents to develop along the surface of the insulator from the energized side to ground. If the leakage currents are high enough, flashover occurs.

3.5.2 Dry Bands

Some contamination and pollutants form **dry bands** (conductive bands) that make very good conductors across the insulator surface. Eventually, leakage current flows across the insulator surface as the voltage literally jumps from dry band to dry band. If enough time passes without cleaning, there will be a short circuit to ground, creating flashover.

3.5.3 Flashover Effects

Flashovers caused by ionized air or pollution do not normally damage an insulator in good condition. When flashovers occur on contaminated or broken insulators, there can be serious damage to equipment, as well as intermittent power outages. They are usually intermittent faults with no apparent pattern and can be difficult to troubleshoot.

GOING GREEN

Ozone

Corona is a problem in heavily populated areas because it produces ozone. Ozone adds to air pollution; thus it is a health hazard. Even at low concentrations, ozone is toxic. Excessive ozone can interfere with lung function and cause irritating respiratory problems.

Ozone is found naturally in small concentrations in the stratosphere, a layer of the upper atmosphere. Ozone in the stratosphere is good ozone because it protects Earth's surface from dangerous ultraviolet rays.

Ozone in the troposphere, the lowest atmospheric layer, is bad ozone. It is manmade, the result of air pollution from burning gasoline and coal. Automobile exhaust and industrial emissions release nitrogen oxide gases (NOx) and volatile organic compounds (VOC). Sunlight causes NOx and VOC to combine chemically with oxygen to form ozone during high-temperature conditions of late spring, summer, and early fall. High levels of ozone occur in afternoon heat and dissipate as nighttime cooling occurs.

3.5.4 Weather Effects

Wind direction aids or retards pollution deposits. An offshore breeze can blow salt deposits onto insulators. Wind can blow industrial pollutants toward the transmission line. Weather conditions of fog, high humidity, or light rain wet the contaminants on the insulator surface. Heavy rain typically washes the pollution layer off the insulator.

In known pollution conditions, composite-type insulators manufactured with **hydrophobic** materials are used. Hydrophobic materials repel water, causing it to form beads that do not run together. This prevents the water from saturating the contaminated surface and creating a large conductive area. Continuous creepage or leakage current paths will not form on the surface of insulators made with hydrophobic materials. This eliminates most flashover damage and reduces much of the need for washing. The most common hydrophobic material is a silicone rubber composite. Hydrophobic composite insulators are lighter in weight than other types and are thus easier for line workers to install.

3.6.0 Washing Insulators

In areas where there is not enough natural precipitation to clean pollution and contaminants from insulators, some method of cleaning is necessary. These methods include the following:

- *Manual cleaning* – Dead line
- *High-pressure washing with water* – Dead line
- *High-pressure washing with water* – Live line

If the electric utility has enough capacity to enable lines to be isolated for maintenance, they may choose to turn the power off. Work crews then climb the transmission structures and clean the insulators with rags and mild abrasives, or use a high-pressure spray washer to remove the contamination. However, many utilities do not have that option. Power lines are very expensive to build and power demands are great, so most utilities do not have the excess capacity to routinely shut down a transmission line. They have no choice but to clean the insulators with the power on.

Contaminated insulators are washed using high-pressure water. Normal tap water contains

On Site

Extenders

Extenders are installed over the tops of insulator strings to protect them from increased conductivity caused by the accumulation of bird droppings. They improve the flashover performance of insulators by reducing the surface electric conductivity, reducing leakage current, and increasing the electric strength of the insulators. Bird droppings are slippery and a possible health hazard. The cover protects the insulator from contamination and prevents the kind of situation shown on the unprotected insulator.

81204-11_SA02.EPS

81204-11_SA01.EPS

impurities, which can be conductive or nonconductive. If the impurities are conductive, the water is conductive and creates a hazard. The water must have a minimum resistance of 16.39K ohms per cubic inch (1K ohms per cubic centimeter). Fortunately, most public utility water supplies satisfy that requirement. If the water resistance exceeds the minimum, it cannot be used. If there is no readily available source of nonconductive water, the impurities in the water must be filtered using reverse osmosis or some other distilling method.

All the spray equipment must be electrically bonded and ungrounded. The hose nozzle must be grounded, either to the transmission structure, neutral conductor, or a static wire. A high water pressure is maintained to prevent a solid stream from flowing. The hose must be kept at a safe distance, usually 10 feet, from the insulator. Start washing the insulator from the bottom. This prevents water from dripping onto a contaminated insulator. When a contaminated insulator gets wet, there is a lower resistance on the surface of the insulator. Leakage currents can flow along the surface of the insulator and flashover could occur.

Insulator washing equipment is generally custom-manufactured to meet the customer's specifications. Some companies want their equipment fitted onto an aerial platform so that their line workers can wash from the bucket, some of which can be as much as 200 feet high. Others have their workers climb the transmission structures, so they use a trailer or truck chassis to carry tanks of water. Both of these methods can be time consuming, and they are only possible where relatively easy access to the transmission structures is possible.

In remote areas or in rough terrain, the washing equipment is mounted on helicopters. Washing may take place from a platform mounted on the side of the helicopter or a permanent system installed in the helicopter. These systems have a self-contained washing fluid tank and pump assembly in the helicopter to provide high-pressure washing fluid through an articulating boom. While hovering next to the high-voltage power lines, an operator directs the spray onto the insulators using an articulating boom or hand nozzle. *Figure 11* shows a helicopter using an articulating boom to wash insulators. Note the water tank on the belly of the helicopter.

3.6.1 Other Cleaning Methods

Greasing with a thin layer of silicone grease is very effective. The grease causes the insulator surface to become hydrophobic. Particles on the

81204-11_F11.EPS

Figure 11 Washing insulators from a helicopter.

insulator are encapsulated by the grease and the surface is protected from moisture. The problem with greasing is that it is a time consuming procedure repeated annually and work crews have to remove the old grease before new grease is applied.

3.7.0 Changing Insulators

Changing insulators may be the most frequent live-line maintenance work on transmission lines. The most common method of changing insulators on transmission structures is the strain carrier method. Procedures may vary slightly depending on tools used, insulator mounting, and transmission structure design. Though the general procedure is similar for all insulator changeouts, there could be some variation due to company procedures and manufacturer's instructions.

3.7.1 Changing Suspension Insulators

The first step in changing an insulator is removing the weight of the conductor from the insulator. Work crews do this by attaching a strain carrier to the transmission structure support and the conductor to remove the weight (*Figure 12*). The following is a general procedure for changing suspension insulators. Always follow your company's procedure or the procedure provided by the manufacturer.

- Place the strain carrier parallel to the insulator.
- Attach the adjustable end of the carrier to the cross arm or the insulator attaching point. Be sure there is enough slack to tighten the strain carrier in a later step.

81204-11_F12.EPS

Figure 12 Crew changing a suspension insulator.

- Hook the fixed end of the carrier to the conductor.
- Tighten the strain carrier and disconnect the conductor from the insulator.

> **NOTE**
> If the strain carrier is not being used, the remaining bells on the insulator must be supported.

- Use a cotter key puller to remove the cotter key holding the conductor to the insulator. Tighten the strain carrier even more if the cotter key is difficult to remove.
- Position the conductor so it is clear of the remainder of the operation. Remove the old insulator, as well as any hardware that should be changed out.
- Mount the new insulator, replace the cotter key, then move the conductor to the connector on the insulator and tie it in.
- Remove the tension from the strain carrier transferring the weight of the conductor to the insulator, then remove the strain carrier.

3.7.2 Changing Strain Insulators

Strain insulators are in a nearly horizontal position so line workers slide the strain carrier (*Figure 13*) over the insulators toward the conductor. The following is a general procedure for changing strain insulators. Always follow your company's procedure or the procedure provided by the manufacturer.

- Attach the strain carrier to the conductor side of the of the strain clamp.

81204-11_F13.EPS

Figure 13 Replacing a strain insulator.

- Fasten the takeup end of the strain carrier to the transmission structure or cross arm. Be sure there is plenty of slack so the strain carrier can be tightened.
- Position the cradle to hold the insulators.
- Hang the adjustable hook over the conductor. Connect the other end to the transmission structure or cross arm.
- Draw up on the strain carrier to take the load off the insulators. After the load is off the insulator, replace the damaged bells or the entire insulator string as necessary.
- Loosen the strain carrier to return the load to the insulators.
- Remove all tools in the reverse order attached.

3.7.3 Other Method of Changing Insulators

For lighter loads, insulators may be accessible from the bucket. Work crews may use the winch on the bucket in order to raise the conductor to remove the strain from the insulator. See *Figure 14*.

In some applications, work crews use a J-cradle to raise and lower insulators. *Figure 15* shows a J-cradle in use. Notice there is no strain carrier in this operation. The remaining three insulators take up all the conductor strain.

3.7.4 Corona Rings

Early versions of corona rings were often installed with the wrong orientation. Inspect corona rings

81204-11_F14.EPS

Figure 14 Removing strain on an insulator with the winch on the bucket.

81204-11_F15.EPS

Figure 15 Insulator replacement using a J-cradle.

for correct mounting, carbon deposits, and pitting from flashover. *Figure 16* shows a bundle of large strain insulators with corona rings installed.

4.0.0 Conductor Inspection and Maintenance

There are more than 300,000 miles of transmission line in North America. Much of it was built between World War II and the mid-1970s. A very large amount of the currently installed aluminum conductors are 40 to 70 years old. The aluminum conductors installed during that period had a projected lifespan of 30 years, with a maximum operating temperature of 158°F. All those older conductors are well beyond their expected service life.

Conductor fatigue is becoming a major issue on transmission lines. Line workers are responding to calls of broken strands in suspension clamps, frayed splices, and damaged spacers. Corrosion and the weather, along with the normal aging process, are causing increased conductor failures. Higher voltages have increased operating temperatures, which further degrades conductors.

81204-11_F16.EPS

Figure 16 Corona rings on strain insulators.

4.1.0 Conductor Inspections

Inspectors typically find loose, missing, or worn hardware; misaligned components; broken or worn conductors; and damaged overhead ground wires. Vibration dampers can corrode and shift position on the conductor, rendering them ineffective. Spacers may become bent from wind vibration and may also corrode. A corroded spacer can break outright, or it can bend, shortening the distance between conductors. Conductors, like the one shown in *Figure 17*, are often damaged by gunfire. Vandals often attempt to cut conductors and break insulators for target practice.

As conductors age, they lose tensile strength. At high operating temperatures, the weight of the conductor creates a warped appearance that takes on a permanent elongated shape.

Inspectors look for corrosion on aluminum conductors and rust on steel conductors and the overhead ground wire. Old aluminum conductors with a steel core are susceptible to rust. Older copper conductors may be very brittle.

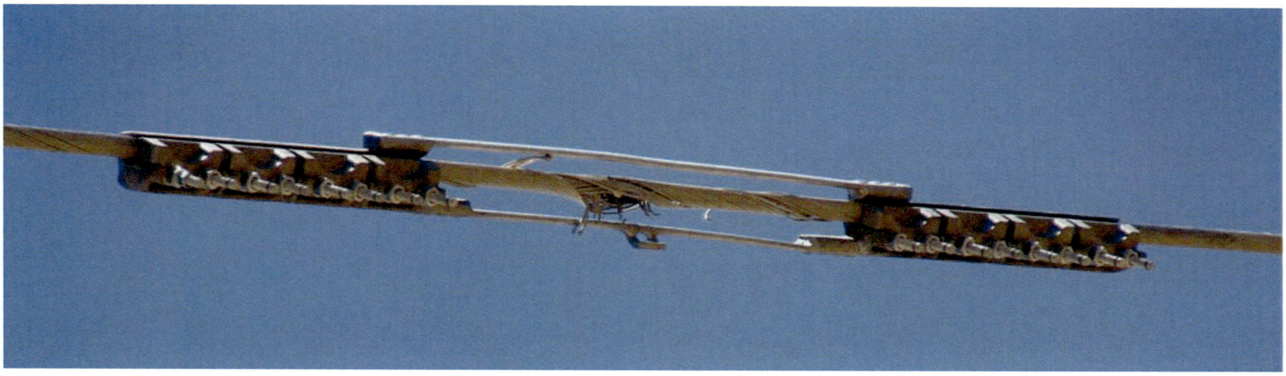

81204-11_F17.EPS

Figure 17 Conductor with gunfire damage.

4.2.0 High Current/High Temperature

Current flow generates heat. Conductor temperatures in excess of 250°F (121°C) are common. For most transmission components, operation at an elevated temperature shortens conductor life. Evaluating the impact of prolonged high-temperature operation on a conductor manufactured 50 or 60 years ago is nearly impossible, since most available test data is for recently manufactured components. Work crews doing conductor inspections have to be aware that corrosion of transmission line components accelerates in a high-temperature environment.

Compression splices are also sensitive to operation at high current levels. Unfortunately, it is impossible to predict the level of deterioration from the effects of time and temperature. Old compression splices made using unknown methods make the problem more difficult to evaluate.

4.3.0 Line Sag

Public safety and government regulatory requirements mandate that conductors remain specific distances above the ground, roads, waterways, structures, and trees, as well as other transmission and distribution equipment. For that reason, conductor sag must be controlled. *Figure 18* shows the sag in a transmission line.

Some sag is natural and expected due to the weight of the conductor. Line sag increases from thermal expansion of the conductor due to high ambient temperatures, low wind speeds, and high line current. Increased demand for power, particularly in the summer when more air conditioning is in use, increases the temperature from current flow. This, in turn, increases the chance for excessive sag.

Transmission line sag may be the most common source of limited line ampacity, which, in turn, is a cause of brownouts. Heating of overhead conductors due to current flow, coupled with a high ambient temperature, creates thermal expansion. This causes conductor sag to increase. This condition is more evident in lines rated from 115kV to 230kV. The amount of sag created by high temperatures is usually about five feet. That may not seem like much, but the increased distance is enough to affect ampacity.

With increased demand for power, some power companies increased the maximum temperature limits of conductors. This solved the immediate problem of customer demand, except on hot summer days. Some conductors (ACAR, AAAC, or AAC) have a normal increase of sag because of creep elongation over time. This effect is greater when factoring in extended operation at high temperatures.

4.3.1 Sag Mitigation

Traditional methods of reducing sag include building a new line to either replace or reduce the load on an existing line; reconductoring with

81204-11_F18.EPS

Figure 18 Conductor sag in a transmission line.

a larger diameter conductor; and bundling more conductors on existing transmission structures. In some cases, power companies add more transmission structures to decrease span length. They might also increase clearance under the span by raising the transmission structures. Some power companies actually derate the line in hot weather. However, it is not always a good procedure because of the increased load demands in hot weather. Each of these methods is expensive and time consuming unless done in conjunction with other transmission line upgrades.

Automatic sag mitigation devices are available for installation on transmission lines. One such device, the SmartConductor, has a shape memory alloy core that shortens when heated. The memory alloy is in a 6-foot length of conductor that workers splice into the conductor near the transmission structure. Another device is the sagging line mitigator (SLiM). This device is spliced anywhere on the conductor and actually becomes part of the conductor. As the line heats up, a bimetallic insert contracts unevenly. The contraction moves a lever which shortens the conductor, reducing sag. Both of these devices are relatively easy to install using common splicing procedures. They only shorten conductor length a few inches, but on an 800- to 1,000-foot span it can increase sag clearance by several feet. An example of the SLiM device is shown in *Figure 19*.

81204-11_F19.EPS

Figure 19 SLiM sag mitigation device.

4.4.0 Overhead Ground Wires

Inspecting overhead ground wires (OHGW) is much like inspecting conductors. Replacing overhead ground wires is simpler than replacing current-carrying conductors since they do not require the same careful handling. Most OHGW is stranded steel.

4.4.1 Fiber Optic Overhead Ground Wires

When stringing optical ground wire (OPGW), travelers and blocks must have a protective lining to prevent crushing of the optical fibers. A swivel is generally used on the pulling line, along with an anti-rotation device, in order to prevent twisting. The slope of the cable should never be steeper than three to one (three horizontal and one vertical). If terrain or obstructions prevent placing the tensioner in the proper location, the first sheave is lowered to obtain the three-to-one slope, and then raised during hardware installation.

A typical stringing tension is in the range of 15 percent or less of the OPGW-rated breaking strength. The maximum stringing tension should be limited to 20 percent of the rated breaking strength. Pulling speeds should be in the range of 1.2 to 3.1 mph (2 to 5 km/h). Always consult the manufacturer's recommendations before stringing OPGW.

The sagging method for OPGW is similar to that used for conductors and OHGW. Sag and tension information is supplied by the OPGW manufacturer. The greatest concern of line crews is not to crush the fiber optic strands. The grips must be specifically designed for OPGW. Manufacturers provide closely machined grips designed to match the diameter of the OPGW cable. With some OPGW cables, line crews can use formed wire grips if the manufacturer approves them. Unlike conventional conductors, OPGW cables are spliced at the transmission structure, rather than mid-span.

4.5.0 Connector Inspection

Inspect connectors and other hardware for roughness. Ensure that the conductor is secure and they are tight on the insulator. When an infrared inspection reveals a hot connector, a common practice is to install a shunt around the connector. This provides a parallel path for current flow so there is less current flow through the hot connector.

4.6.0 Splices

Line crews insert splices at mid-span or near the point of tension at the transmission structure. The workers in *Figure 20* are repairing a mid-span splice. Span length on transmission lines is long, so it is important that splice joints have great mechanical strength. Of equal importance is that splices must have a very low electrical resistance.

When installing a splice, workers must not place themselves between the ends of two conductors or in series to ground with either conductor.

81204-11_F20.EPS

Figure 20 Transmission workers repairing a splice.

If the splice is in mid-span, ground grids must be installed at the pull and tension sites. The conductor cart and all other mobile equipment, tractors, reel trucks, lift trucks, and anything that may be holding the conductor ends must be bonded to the ground grid. The two conductors are bonded to a common ground connected within 10 feet of the conductor ends. A shunt jumper is placed across the opening between the ends of the two conductors. All work must be performed within the perimeter of the grid.

Work crews may use a metallic grounding mat. In this case, the mat and the conductor ends must have a common ground. A jumper is installed between the conductor ends and all ground rods are bonded together. If not using either the grid or the mat, workers must install the splice work from an insulated platform or from a helicopter.

There is a three-step testing method to determine the condition of a splice:

- Measure the wire resistance immediately adjacent to the splice.
- Measure the resistance of the junction between the wire and one end of the splice.
- Measure the resistance between the wire and the other end of the splice.

The following is a general procedure for preparing to install a splice. Always follow your company's procedure or the manufacturer's instructions.

- Attach two wire grips facing each other approximately 12 to 20 feet on either side of the damage.
- Attach a chain hoist to the grips and take up tension. As the tension increases, slack will appear in the damaged component conductor.
- Continue to increase tension until there is enough slack to make repairs.

NOTE

It may be necessary to cut the damaged conductor before installing a splice. To prevent any possibility of the conductor span falling to the ground, attach a second set of grips and apply tension with the second hoist to the damaged conductor before cutting it.

4.6.1 Clamp Splices

A clamp-type splice is an economical, quick-to-install splice. It is a shunt-type device, designed for installation on a live conductor or over an existing connector. It restores both electrical and mechanical integrity to the conductor. Line crews can easily install clamp splices with a hot stick. Clamp splices are also useful for doing repairs at suspension clamps and at some damaged spacers that have severe fretting and broken strands. The clamp splice is positioned across the damaged area with keepers open. Once in position, line workers tighten the bolts with a hot stick. *Figure 21* shows line workers installing a clamp splice.

Larger clamp splice assemblies have a flexible frame comprised of sections of conductor to allow the splice to flex to various degrees of bend, and to accommodate vibration issues. On extra high-voltage lines, workers install corona rings across the clamping bolts.

81204-11_F21.EPS

Figure 21 Line workers installing a clamp splice around the damage on a conductor.

4.6.2 Compression Splices

When making a compression splice, make sure all dies are the correct size. Always follow the sequence specified by the manufacturer while making the presses. That holds true for any joint compounds added to prevent corrosion, improve electrical contact, or aid in mechanical holding. Not following the proper procedures results in a faulty splice. Faulty splices are a potential hazard and may lead to an outage. The following is a general procedure for performing a compression splice. Always follow your company's procedure or the manufacturer's instructions.

- Place the aluminum splice body on one end of the conductor. Slide it far enough for there to be room for the steel sleeve.
- Cut the aluminum strands off one-half the length of the steel sleeve plus one half-inch more. Make the ends square with the conductor.
- Clean the exposed steel surface. Insert the steel core no further than halfway into the sleeve. Be sure to correctly center the joints on the conductor.
- Compress the sleeve onto the steel. Make the first compression on the center to squeeze against both ends of the conductor. Continue applying compression, moving toward the end of the sleeve.
- Clean the aluminum strands. Unwrap them if necessary to remove all residue. Rewrap the strands and slide the aluminum body over the sleeve.
- Insert joint compound into the filler holes on the sleeve until it begins to flow out between the conductor and the aluminum body.
- Press the aluminum sleeve, starting in the middle and moving toward the ends.

4.6.3 Automatic Splices

Automatic splices are often used to make quick temporary repairs to conductors. To install an automatic splice, start by selecting the proper size splice according to the type of conductor. The following is a general procedure for installing a compression splice. Always follow your company's procedure or the manufacturer's instructions.

- Cut the conductor, making sure the ends are straight and square with the conductor.
- Clean the conductor, but do not apply any inhibitor.
- Insert the conductor into the pilot cup with a single smooth motion until it strikes the stop in the center. If the conductor does not go in all the way, do not twist or force it. Remove it, trim the strands, and start over with a new splice.
- Set the conductor in place by giving the conductor a quick pull.

CAUTION

Never reuse an automatic splice. Avoid using automatic splices in coastal areas where salt corrosion is a problem. Do not use automatic splices in slack spans, as jumpers, or when stringing conductors.

4.7.0 Spacers

Spacers are used to separate conductors on bundled lines with multiple conductors per phase. Two and four conductors placed next to each other in a bundle is a common arrangement. Occasionally there are six conductors in a bundle.

Spacers are subject to bending fatigue caused by wind damage, and commonly require replacement. Spacers are replaced using a high bucket truck; by placing personnel on the conductors in a conductor cart; or from a helicopter. The worker in *Figure 22* is installing a spacer near the transmission structure. The location and manner of installation varies with the type of spacer and the manufacturer. Always install spacers according to the manufacturer's instructions. When reconductoring, install spacers immediately after clipping-in the conductor.

When unequal spacing is used, accuracy in spacer location is critical to protect the line against damage. Use ground targets or conductor cart distance counters to accurately locate the spacers. Verify cart footage counter accuracy to span length to ensure correct spacer positioning.

81204-11_F22.EPS

Figure 22 Installing a spacer.

4.8.0 Vibration Dampers

Conductors have a natural swinging motion from blowing wind. This wind-induced motion is normal, but there are three types of wind-induced vibration experienced on transmission lines that can be harmful. If left unchecked, they can result in enough reduction in electrical clearance to cause flashover.

- Aeolian vibrations occur at frequencies from 5 to 60 Hz, and are caused by low-velocity winds. If Aeolian vibrations are large enough, they can lead to conductor fatigue and other transmission line components. Engineers use several methods reduce the amplitude of Aeolian vibrations in transmission line design. They specify reducing the conductor tension or the use of conductors with physical characteristics that provide natural damping. Line crews add vibration dampers when they string a conductor.
- Galloping vibration is a leading cause of outages. This phenomenon occurs when strong winds blow across ice-covered power lines causing violent bouncing. The conductors can strike each other, resulting in outages. Galloping lines also break cross arms, insulators, and transmission structure sections. Conductors sometimes break, fall to the ground, and cause outages.
- Wake-induced vibrations occur on lines with bundled conductors. They are caused by aerodynamic forces on the downwind conductor of the bundle as it moves in and out of the wake of the upstream conductor. Tilting the bundles and ensuring a large conductor spacing in the bundle keeps this vibration at acceptable levels.

If dampers are necessary to prevent conductor damage from winds, it is best to install them immediately following clipping. Damage can occur in a matter of a few hours from wind damage at initial conductor tensions. *Figure 23* shows two line workers in the process of installing vibration dampers.

Line workers may install vibration recorders to measure vibrations on lines in operation. They record ambient temperature and the wind velocity acting on the conductor. Engineers can use the data to assess line conditions and to schedule maintenance.

81204-11_F23.EPS

Figure 23 Transmission line workers installing vibration dampers.

81204-11_F24.EPS

Figure 24 A conductor cart.

4.9.0 Conductor Cart

A conductor cart (*Figure 24*), sometimes called a spacer cart, is a device used by workers to ride on bundled conductors. It enables them to inspect the conductors for damage and to install spacers and dampers where required. Conductor carts may be manually operated or powered. Line workers ride manual conductor carts like a bicycle. Manual carts are limited to use in flat terrain. They are difficult to pedal on long spans or where the span goes across a mountain. Hydraulically operated carts are more common. In some cases, ground workers pull some carts along the span.

Line crews use conductor carts for inspections, general maintenance, and attaching spacers and dampers after reconductoring. In *Figure 25*, line workers are installing spacers on all three spans after reconductoring.

81204-11_F25.EPS

Figure 25 Line workers installing spacers.

Exercise care with conductor carts to ensure that the combined weight of the cart, line worker, and equipment does not increase the sag to the point where it cannot clear obstructions below the conductor span. To prevent overloading, ground crews may deliver parts and tools to the worker in the conductor cart using a bucket. If the cart is too high, or the terrain is too rough, helicopters are used to deliver supplies. Either the helicopter lowers supplies by hoist, or a crew member hovering next to the cart hands them to the worker.

NOTE

The load of the worker, car, and equipment should be equally distributed to all subconductors of the phase. This is particularly important when attaching spacers.

4.9.1 Armor Rods

Armor rods are commonly referred to as helical rods. Armor rods are used to reduce bending stress in a conductor and to strengthen a conductor at a location where it would be subject to chafing and vibration.

Figure 26 shows a line worker installing armor rods. When installing armor rods, hold the rod firmly in position against the conductor, then begin wrapping the rod around the conductor, working from the center toward one end. Then return to the center and wrap toward the other end. Be careful not to let the rod slip into the conductor strands, as this makes the remaining rods difficult to position. Apply the remaining rods the same way. Be sure to align the rods using the center color mark on each rod (or rod subset). Do not let rods overlap an adjacent rod.

If the conductor is flexible cable, some tension may be needed to stiffen the conductor. The helical rods are easier to install on a stiffer conductor. Be very careful during handling so as not to put a permanent bend in the helical rods.

4.10.0 Conductor Monitoring Equipment

The development of sophisticated electronic testing devices enables power companies to monitor transmission line performance in a number of ways. New technology has made possible the real-time monitoring of sag and overhead line

81204-11_F26.EPS

Figure 26 Installing armor rods.

temperatures. The cost and installation of monitoring equipment is low compared to that of an outage or major equipment replacement. It is cost effective in preventing outages from unscheduled maintenance. Most power companies routinely track real-time values of bus voltage and circuit power through current and voltage monitors.

Real-time monitoring of lines allows the detection of imminent failure events. By performing an ongoing comparison of calculated and measured sag and temperatures, problems can be detected before damage that leads to an outage occurs. Engineers note the variation in conductor tension sag in relation to temperature. Evaluation of the data, often done with computers, permits remote detection of conductor steel damage or structure support failures. Computers monitoring the lines can often detect faults as soon as they occur.

4.10.1 Infrared Thermovision Testing

Inspectors and maintenance crews use infrared thermovision cameras that detect heat. The thermovision cameras detect hot spots that require maintenance. The inspection can be done from a helicopter, specially equipped trucks, or by a line worker on the ground. Problem areas show up as a bright white image on a black and white thermovision camera. On a color camera, hot spots change from black at low temperatures through red, orange, and yellow, to white at the highest temperature. *Figure 27* shows an infrared photo of corona taken with a handheld camera. The small white objects are corona hot spots.

Some helicopter-mounted thermovision cameras are equipped with global information system (GIS) technology. As the inspection progresses, there is a continuous record of fault location, what was found, and where it was found.

5.0.0 Tower and Pole Inspection

Many utilities conduct a climbing inspection one year after installing transmission lines. Inspections generally occur every five years after the initial climbing inspection. They typically alternate between a walking inspection five years later and another climbing inspection five years after that.

Aerial inspections are often performed using helicopters. The inspector may stay in the helicopter or the helicopter may hover close enough for the line worker to step onto the transmission structure and move about to conduct the inspection. Some helicopter pilots are trained in transmission line construction and maintenance practices. They can fly line survey inspections without the need for an inspector as a passenger.

Helicopters are faster for transmission structure inspections and for surveying conductor and connector hot spots. Nevertheless, a ground inspection is better for some problems, such as finding danger trees, erosion, footer damage, loose or missing nuts on anchor bolts, corrosion below the ground, damaged guys, and broken ground wires.

On walking inspections, the line workers often carry test equipment to evaluate transmission line performance. The equipment can detect corona discharges, RFI, and defective equipment.

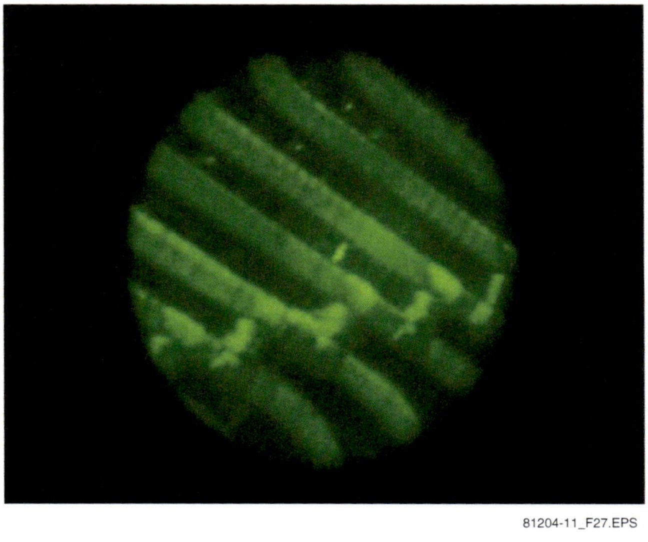

81204-11_F27.EPS

Figure 27 Infrared photo of an insulator with corona.

Stabilized binoculars are common in transmission structure inspections. They are used by both ground and airborne inspectors. They are useful because the inspector has a steady view of transmission structure equipment, uncluttered by movement.

The inspector submits a report of his or her findings in enough detail so that engineers and work planners can create a useful work plan.

5.1.0 What Inspectors Look For

There is some overlap between a climbing inspection and a walking inspection, and inspectors check some items in both inspections. On an aerial inspection, whether from a helicopter or an actual climb, an inspector typically looks for the following problems:

- Broken insulators
- Obvious contamination on insulators
- Broken cross arms
- Excessive line sag
- Corroded vibration dampers and spacers
- Broken and corroded connectors
- General rust and corrosion
- Damaged conductors and splices
- Broken ground wires
- Nests of large birds
- Signs of structure vibration and misalignment
- Ground erosion
- Aircraft warning light operable (if installed)
- Pole damage from wildfires
- Split and cracked poles near the ground from freezing and thawing
- Leaning poles from frost jacking
- Anti-climbing devices in place

When doing a walking inspection, an inspector looks for as many as possible of the climbing inspection items just listed, as well as the following:

- Guys anchors that are pulling out
- Guy wire conditions
- Missing guy guards
- Ground wire connections and conditions
- Ground resistance at each structure
- Structure footings for washouts and damage
- Washout and surface erosion on slopes.
- Danger trees and tall brush
- A buffer zone of vegetation in any right-of-way located within a few feet of water bodies.

5.1.1 Transmission Structure Grounding

Transmission structures must have a very low resistance to ground of 10 ohms or less. A low resistance limits the number of possible outages from lightning strikes. A low resistance will minimize flashover across the insulators when lightning current discharges to ground. Ground connections should be inspected for tightness and freedom of corrosion. All ground cables must be firmly mounted.

5.2.0 Wood Pole Inspection

Maintenance required on wood structures and cross arms in a transmission system is minimal. A wood pole lasts for many years; 20 years or more is not unusual. Common problems are the need to straighten poles that have settled unevenly into the ground, and wood shrinkage that causes hardware to become loose. Cross-arm brackets and mounts should be tight and secure. Bolts should not wobble in the holes because of pole shrinkage.

Inspectors look for damage and leaning poles caused by soft soil. Construction near the right-of-way may have changed the landscape since the original installation. The inspector must be observant to notice whether changes have created a hazardous location for the transmission structure.

Poles should not lean more than 5 degrees. To correct excessive lean, either replace the pole or add support by installing another pole alongside.

Assume the pole is dangerous to climb if there are numerous woodpecker holes, knots, or burn marks at about the same height. These conditions would indicate a weak spot on the pole. *Figure 28* shows a pole seriously weakened by woodpecker damage. Notice that the protective wrapping did not deter the woodpeckers.

5.2.1 Pole Decay

Wood poles receive a treatment with preservatives to prevent decay, but small organisms, insects, and fungi can cause wood preservatives to break down.

81204-11_F28.EPS

Figure 28 Woodpecker damage.

On occasion, work crews bore a pole to determine the presence of internal voids. Poles with internal decay can be treated with insecticides. External decay can be removed, and the area treated with preservatives and wrapped with a moisture-proof barrier. Poles weakened excessively by internal or external decay, like the one in *Figure 29*, must be reinforced or replaced. That pole appeared strong, but the conductors were holding it in position. As soon as work crews removed conductor strain, the pole collapsed. It is possible to extend the life of a pole by regular inspections and prompt treatment of damage and decay.

81204-11_F29.EPS

Figure 29 Wood pole rotted from the inside.

5.2.2 Inspecting for Pole Decay

Rot commonly occurs where the pole enters the ground (*Figure 30*). Check for ground area pole rot by driving a screwdriver into the pole with a hammer at about six inches below ground. If the pole is good, the screwdriver will meet with immediate resistance and will not pierce the pole. A rotted pole not offer any resistance. The screwdriver may pierce anywhere from several inches to all the way through the pole.

An inspection normally includes sounding the pole by hitting it with a hammer at ground level to approximately six feet above ground and listening for a hollow sound. This method exposes defects from rot, birds, insects, or shelling (separation of the pole's outer shell). Insect colonies can eat away at a pole until it no longer supports its own weight.

Nondestructive testing of wood poles is possible using test equipment developed for that purpose. The line worker enters the wood type and pole diameter, then strikes the pole with a steel ball. A microprocessor analyzes the vibrations. The test equipment manufacturer's instructions must be followed to ensure an accurate test.

5.2.3 Old Poles

If a pole is older than 20 years, age-related problems might be evident. Inspectors and line work-

81204-11_F30.EPS

Figure 30 Pole rotted at ground level.

ers must beware if the wood species has a history of failure. Soil that is soft, loose, or wet could mean the pole is not firmly embedded in the ground. Nearby trenches from erosion may have loosened the soil holding the pole. Look for shallow settling indicated by the markings of a former ground line showing above the current ground level. During maintenance, support weakened poles with a line truck boom, temporary or permanent guys, or by lashing another pole alongside the weak pole.

5.3.0 Steel Structure Inspection

A steel structure lasts for many years, so very little maintenance is necessary on steel structures and cross arms. A common problem is a structure footing settling unevenly into the ground. Another is gunfire damage on the small components at the tops of transmission structures. Inspectors also look for damage and leaning structures caused by soft soil. Construction near the right-of-way may have changed the landscape since the original installation. The inspector must be observant to changes that may have created a hazardous location for the structure.

Work crews examine the physical construction of the structure. They look for wind and ice damage such as bent, loose, or missing structural members and verify that all climbing aids, platforms, and catwalks are secure. They also look for loose and missing bolts. It is sometimes necessary to replace the structure. *Figure 31* shows work crews replacing a structure.

Bolts and nuts that hold steel plate and lattice members together must be checked for corrosion and tightness. Bolt ends near conductors must be covered if evidence of corona is present.

5.3.1 Structure Finish Inspection

Inspect the tower finish for rust and corrosion. If the structure is painted, note the condition of the paint and whether any rust is showing. Check coatings for pitting and bare spots, as these are the points where rust and corrosion begin. If rust is evident, there is usually more under the coating. Remove any paint or coating in an area at least two feet around the rust. Look for water collection in structural members. Unplug or add drain holes to remove trapped water.

If the structure is near an airport, verify that FAA color marking is still in good condition with a clear contrast between white and orange.

81204-11_F31.EPS

Figure 31 Replacing a structure that has a bent frame.

5.3.2 Foundation Inspection

When inspecting the structure base foundation, observe the ground conditions for settlement or movement, as well as erosion. Observe the site condition for standing water, poor drainage, high brush, and danger trees. On the structure base, make sure all nuts and lock nuts are tight.

Check the condition of the concrete for crackling, **spalling**, or splitting. Kick or strike concrete footings to see if the concrete crumbles. Look for chipped or broken concrete, a honeycombing appearance, and low spots that collect moisture.

5.4.0 Other Pole Types

Fiberglass poles are immune to the problems found in wood poles. Inspectors should look for cracks and stress lines around fittings.

Power companies use concrete poles in areas where timber is scarce and concrete is readily obtainable. Concrete poles are immune to insect damage and other forms of decay. Concrete poles are strong due to reinforcing steel, iron, or aluminum mesh. The reinforcement compensates for bending stresses due to wind loads and conductor tension. Concrete poles are particularly useful on a narrow right-of-way as they can be very tall.

Concrete poles are stronger, lighter, more durable, and very cost effective. Inspectors have to be alert for cracks that might admit moisture. When water freezes, it expands and is likely to cause internal structural damage.

Composite poles can be installed with less equipment and manpower than other poles. They significantly speed up recovery from severe weather and natural disasters. Composite poles have a uniform diameter and strength and one pole type is usable in many applications. That feature eliminates the need for storing multiple pole types in the inventory.

Aluminum structures are lightweight and easy to install. On a major upgrade, aluminum structures are assembled at a convenient location. Helicopters then airlift a complete transmission structure with stringing sheaves, insulators, and tag lines. In short, everything necessary is quickly assembled on the ground without having to climb. Aluminum structures do not require painting. In coastal areas, inspectors must look for corrosion from salt in the air.

5.5.0 Guy Wires

A transmission structure holds up conductors, but the conductors help secure the transmission structure poles. Guy wires are not common on steel structures, but may be necessary to counteract the forces of a heavy conductor that could pull the structure over. They are commonly used on wood structures (*Figure 32*).

81204-11_F32.EPS

Figure 32 Guy wires on wooden transmission structures.

Guy wires may also be used on dead-end structures to counter the weight of the conductors pulling against the structure. Guy wires compensate for the heavy forces applied to a structure when the transmission line changes direction. Unbalanced forces on transmission structures occur when the conductor size changes or when there are different conductor types and tensions.

Since guy wires are in locations of a potential problem, it is important that workers doing inspections carefully examine them for problems. Guy wires must be kept clean and clear of dirt and debris. Check for corrosion, breaks, nicks, kinks, and other damage. Look for broken strands on the guy wire. Verify that cable clamps are applied properly and the bolts are tight.

Check that any preformed wraps are properly applied and fully wrapped with the sleeve in place. Verify that shackles, bolts, and pins are in place and in good condition.

Use a tension meter to check guy tension according to the manufacturer's recommendations. Surprising amounts of slack can appear without much warning. The tension meter shown in *Figure 33* measures up to 10,000 pounds (4,536 kg). Variations in guy tension from temperature and wind are normal. If the tension is significantly different from specifications, determine the cause immedi-

81204-11_F33.EPS

Figure 33 Guy wire tension meter.

ately. Some possible causes of loose tension may be structure settling, heavy wind or ice loading, anchor movement, footer settling, or connection slippage. Do not check or adjust guy tensions during times of high winds.

5.5.1 Guy Wire Anchors

Most existing guy anchors are hot-dipped galvanized with a zinc coating. Zinc is used to prevent corrosion, but it eventually wears away. Manufacturers apply epoxy coatings to guy anchors, but a coating damaged in shipping or installation can create small defects in the surface. These small pits in the coating are a source of corrosion. If a guy wire or anchor rod has lost its galvanized coating, it could mean that rust or corrosion has weakened the metal.

The resistivity of the soil is an important factor in rate of corrosion of guy anchors. Clay soil, particularly in a wet climate, promotes corrosion. When checking guy wires, dig about 12 inches down around a suspect anchor. A small amount of corrosion in that area indicates the likelihood of corrosion further below the surface. It may be necessary to install a temporary guy and remove a suspect guy anchor. Clean the anchor completely before evaluation, as the products of corrosion leave a thick scale on the anchor. The scale gives an appearance of a normal-size anchor. Workers can determine if a corroded anchor requires replacement only after cleaning.

Measure the anchor with a micrometer or ultrasonic thickness tester. If it is too small, replace it. If the size is acceptable, saturate the anchor with corrosion-inhibiting epoxy before reinstalling it. Be sure to cover the anchor completely, with no gaps in the coating. Take care to make sure there is no damage to the coating during reinstallation.

Encasing the guy anchor in concrete is one way to prevent corrosion. However, if the concrete cracks, water could enter the crack. When the water freezes and thaws, it can damage the concrete and expose the anchor to corrosion.

5.5.2 Ungrounded Guy Wires

Ungrounded guy wires on transmission structures have insulators that must be inspected for integrity. Some old transmission structures may have wood insulators on the guy wires. Wood insulators have arcing horns to provide a bypass path for lightning. Inspect the arcing horns for damage and replace them with fiberglass insulators if necessary.

5.6.0 Transmission Structure Lighting

Some transmission structures near airports have aircraft warning lights installed. Check to ensure that the conduit, junction boxes, and fasteners are weathertight and secure. Make sure all drains and vents are open and free from dirt and debris that tends to accumulate in them. Visually inspect the wiring for cuts, nicks, frays, and loose connections. Verify that the controllers, flashers, photo controllers, and alarms are functioning. Clean light lenses and verify that both the primary and secondary bulbs are working. Many companies replace all the bulbs at once to ensure they are in good condition.

5.7.0 Right-of-Way Inspection and Maintenance

Use of the area below a transmission line is restricted to keep objects away from energized conductors. One reason for this is that overhead lines and structures may shed ice, creating a hazard. Transmission structures originally installed in remote locations may no longer be remote, as developers move further into rural areas. Look for evidence of development that can cause problems. In urban areas, the public often takes dangerous risks near transmission lines. They fly kites and balloons near the conductors, for example. Nearby businesses sometimes encroach onto the right-of-way with buildings and machinery.

Always check the grading on the right-of-way for signs of erosion and construction debris that can weaken a transmission structure. Nearby construction of buildings and roads can change the drainage characteristics of a right-of-way. The resultant soil erosion may weaken transmission structures and anchor guy footings. Eroded areas must be filled and graded to restore good drainage.

5.7.1 Control of Vegetation

Look for and remove trees that could be hazardous to the line of transmission structures. Use small trees and plants for a transition strip from natural ground cover to larger growth. When the transmission line departs from a meadow or other open area, feather the trees in at the entrance into densely wooded areas for a distance of 150 to 200 yards. Land on either side of the center line must be clear of any brush or small trees that could grow to reach the conductors.

Remove all trees, stumps, brush, and other flammable material except grass. Cut all vegetation three inches high or less to permit the passage of trucks and other maintenance vehicles. If on a hill, measure the height from the uphill side. In some areas, taller vegetation is allowed for environmental or aesthetic reasons. Consult with engineering to see if those conditions apply.

The preferred disposal method for debris produced by clearing is chipping it and spreading it or hauling it away to a designated disposal area. Burning may be acceptable, but must be done in accordance with environmental regulations and all state and local laws. Do not burn the debris if the wind could create a nuisance by blowing the smoke into residential areas or if there is potential to start a grass or forest fire.

WARNING!

Transmission lines often travel through remote areas with limited access. Always check the weather before burning. On high-temperature, low-humidity summer days, the danger of a brush fire raging out of control is very high. Never burn when there are high winds.

The right-of-way is generally treated with growth retardants to prevent the growth of trees, brush, and other unwanted growth. The chemical treatments are carefully selected for effectiveness, depending on the season and the ecological needs of the area. Bare ground caused by natural erosion or maintenance activity must be covered or reseeded.

6.0.0 Switching, Tagging, and Clearance Procedures

Because of power demand, it is not always feasible to shut down a transmission line. For that reason, transmission line inspection and maintenance work is frequently done on live lines. However, the design of the power grid does make it possible to reroute power around some lines, allowing them to be shut down for maintenance.

Every utility has strict clearance procedures, also known as **isolation guarantees** or Do Not Operate Permits (DNOP), to disable the line and to ensure that it is not inadvertently energized while work is being performed. Once established, an isolation guarantee remains in effect until it is surrendered by the field force (generally the maintenance supervisor or crew leader).

In all clearance activities, there is a controlling authority responsible for granting permission to perform work. For purposes of this discussion, this authority can be called the system outage manager. The explanation that follows is generalized. Every company has its own policies and procedures for generating an isolation guarantee and **switching order**.

6.1.0 Application Process

Once it is determined that work is necessary, an application is made to the system outage manager to isolate the circuit. This is done according to company policy and procedures. The person who files the application must have the authority and responsibility to ensure that the circuit remains isolated and grounded throughout the entire work process.

The application must be filed far enough in advance for the system outage manager to evaluate the request. This gives the outage manager time to determine anticipated load demands on the system and assess whether the request affects other maintenance activities.

When preparing an isolation guarantee, the circuit must be identified with a number or easily understood and recognized name. Every device to be de-energized is identified and located on a schematic diagram, an operating map, or a one-line drawing.

6.2.0 Switching Order

After the system outage manager approves the clearance or isolation guarantee, a switching order is issued. The switching order is generally a computer-generated document that specifies what switches to operate and in what order. It

On Site

Air Traffic

FAA Regulation AC70/7460-1K covers power lines near airports to protect aircraft from flying into an unseen hazard. *AC70/7460-1K* states that power line encroachment is the responsibility of the utility company, not the airport. Near airports it is essential to observe clearance above the line as well as below. Watch the planes take off and land to see if there is adequate clearance. Clearance above the line may be a problem if runways are extended or new ones are added.

contains the circuit to be isolated, the purpose for the work, and identifies all switchgear that must be operated. If additional pages are needed, they are numbered to ensure that all pages are accounted for. A missing page could result in a fatal mistake.

The system outage manager designates a person to coordinate the switching order with all parties involved. With some power companies, the switching order is so specific that it controls the placement of the protective grounds. Other companies permit the field force to control the placement of grounds.

6.3.0 Operating Switchgear

Immediately before the start of a planned switching order activity, all parties involved participate in a safety meeting to thoroughly review the job requirements. The responsible field force establishes direct communication with the system outage manager. The switching commences when line crews are ready to start work. The responsible field force is the only person to communicate the switching order. Tags and locks are placed on the affected switchgear in accordance with company procedure. This alerts others that work is underway and prevents inadvertent energizing. After the switchgear is switched to the open position, it must be visually and physically checked to ensure it is off. In some systems, it may be necessary to disable a motor-operated disconnect or open another disconnecting device in series with the specified switchgear.

All switching requests must match the approved switching order. If any deviation from the switching order becomes necessary, all switching activities cease. The field force must reconvene the team to agree on changes before continuing. The field force logs all actions on the switching order, including time of occurrence.

Once the switchgear is off, the system outage manager signs a document certifying that the circuit is isolated and will remain isolated until the field force surrenders the isolation guarantee. Many power companies assign a serial number to the isolation guarantee to distinguish it from all other work in progress.

The crew must make potential checks to verify that there is no voltage on the circuit before beginning maintenance work. At this point, work crews have the authority to install protective grounds.

Upon completion of the maintenance, the switchgear is operated to restore the circuit to operation. The field force then signs the switching order and states the condition of the circuit.

6.4.0 Verbal Instructions

In some instances, particularly when the work site is remote from the system outage manager, the switching order instructions are given over the radio using a three-way process. In this event, the substation worker receiving the switching order writes down each step and then repeats every step in the message back to the system outage manager.

6.5.0 Software-Driven Switch Orders

There are software systems that permit engineers to create the isolation guarantee on an Internet-based system. Once the isolation guarantee is approved, the switching order is automatically generated. When ready to begin work, the switchgear operations are recorded online. Workers at the job site can go online and verify the switchgear settings. The person who has work authority can monitor the process from a laptop.

Once a switching order is in effect, software permits the system outage manager to validate it against operating conditions using the interface to the network analysis tools. This allows for pre-verification of the switching order before any work begins.

In addition to generating switching orders, some software-generated switching orders have the capability of transmitting automatic control signals via **Supervisory Control and Data Acquisition (SCADA)**. These systems remotely operate the switchgear, verify position, and prevent inadvertent closing.

Some software systems permit work to be scheduled months in advance. They automatically generate a switching order and issue it when the work is scheduled to begin.

7.0.0 Inspection Reports

When inspectors locate a fault, there has to be a clear method of reporting the fault that supervisors and engineers can easily understand. It is important that fault reports be understandable and provide a clear description of the problem. Most utilities have standardized maintenance procedures and forms to ensure that fault reporting is clear.

When filing a report, avoid nonstandard codes and shorthand notations on inspection forms. Use standard phrasing and company terms to describe findings. Uniform reporting procedures simplify fault reporting and ensure that supervisors, engineers, and management understand what the inspector finds.

Whether on a routine maintenance task or involved in an inspection, line workers must report any hazardous situations, safety violations, or conditions that could lead to an outage so that repairs can made.

7.1.0 Inspection Forms

Power company inspection procedures usually include the use of standardized forms. Many companies have serialized forms. Assigned inspections, reporting of faults, repairs, and switch orders all use the inspection form serial number. This simplifies fault tracking, maintenance tasks, and record keeping.

Table 1 is an example of an inspection form for an overhead line. Some items on this form are not found on most transmission lines. These items (transformers, sectionalizers, and regulators) receive inspections when there is switchyard equipment associated with a specific transmission line or when a subtransmission line connects with a main transmission line. Disregard items that do not concern the transmission line under inspection.

Place an X in the (U) column for Unsatisfactory or (S) column for Satisfactory, as appropriate. Leave the box blank if the item listed at the top of the column is not present. When an explanation is necessary, place a number in the appropriate column and provide details in a numbered list under Explanation for a Unsatisfactory (U) Rating. It may be necessary to add sheets to the form to explain everything found on an inspection.

The following are examples of criteria that the inspectors might look for within the categories listed on the form.

- *Right-of-way* – In individual spans, note vertical clearances, danger trees, overhanging limbs, and erosion.
- *Pole or structure* – Make a visual inspection at the ground line and above ground for damage or deterioration. Inspect concrete foundations for freeze damage. Note noncompliance with codes, including unauthorized attachments.
- *Pole-top assembly* – Note tilted, loose, rotten, or split cross arms; chipped or broken insulators; and loose, corroded, broken, missing, or misapplied hardware. Look for excessive rust and corrosion on steel towers.
- *Anchors and guys* – Consider location, slackness, clearance, bonding, mechanical damage, and corrosion. If required, note the presence of a guy marker.
- *Conductors* – Note clearance, sag, and conductor damage.
- *Overhead ground wire* – Note clearance, sag, condition of clamps, and conductor damage. Verify that any aircraft obstruction balls are secure. Look for nests of large birds of prey.
- *Line regulator and distribution transformer, oil circuit recloser, sectionalizer* – Most transmission lines do not have these components. Inspect them only if present at a switchyard or at a junction with a subtransmission line. Note plumb and mounting, as well as electrical connections, tank, bushings, jumpers, grounding, and lightning protection. Look for corrosion, damaged bushings, and a properly grounded handle.

Table 2 is another example of an inspection form. It covers specific equipment on a transmission line, such as transmission structures, conductors, guys, cross arms, and right-of-way. It also lists equipment that would only be encountered at a switchyard or subtransmission line junction, and also requires careful entry of the structure or pole number (*Figure 34*).

Some companies prefer this type of form or a variation of it because it has columns for specific faults instead of relying on the inspector's notes and explanations.

Inspectors check for the same things when using this form, except they put an X in the column to indicate a problem. They write an explanation in the notes column. Comment sheets are added if needed to explain problems found during an inspection.

7.2.0 Computer Reporting

Advances in computer technology and software have made very sophisticated and elaborate inspection reporting possible. A software user interface mimics the paper inspection forms that transmission line inspectors and maintenance personnel used for many years. Inspectors can enter data in a laptop or a handheld device as they travel the line.

All the information on the transmission line is in a database, including transmission structure location, transmission structure equipment, conductor type and rating, insulator information, and terrain. Weather tendencies that would affect vibration and spacers, as well as environmental concerns for the transmission line route, are in the database. The database contains all previous maintenance activities to make problem history available to engineers and line workers.

Line crews now have direct Internet access to all the information concerning the transmission line on a laptop computer or handheld device. A moving map display identifies the inspector location

Table 1 Example of an Overhead Line Inspection Form

OVERHEAD LINE INSPECTION FORM	LINE NAME										PLACE AN "X" IN ONE ☐ 1 PH. ☐ V PH. ☐ 3 PH.										NAME OF INSPECTOR
	MAP NO.				SECTION NO.						NOMINAL WIDTH, CLEARED R/O/W										LINE SAMPLE NO. / DATE PREPARED
STRUCTURE LOCATION	RIGHT-OF-WAY		POLE (or STRUCTURE)		POLE-TOP ASSEMBLY		ANCHOR & GUY		PRIMARY CONDUCTOR		SECONDARY & SERVICE		OVERHEAD GROUND WIRE		LINE REGULATOR		DISTRIBUTION TRANSFORMER		SECTIONALIZING DEVICE		**EXPLANATION FOR AN UNSATISFACTORY (U) RATING** *(If additional space is needed, attach a second sheet; include structure location on second sheet.)*
	S	U	S	U	S	U	S	U	S	U	S	U	S	U	S	U	S	U	S	U	
TOTAL DEFICIENT																					
TOTAL INSPECTED																					

Table 2 Transmission Line Inspection Form

TRANSMISSION LINE INSPECTION AND MAINTENANCE LOG SHEET

LINE OR SECTION NO. ____________ MAP NO. ____________

INSPECTOR ____________ DATE ____________

INSPECTOR ____________ DATE ____________

	STRUCTURE NUMBER
STRUCTURE	DATE TREATED
STRUCTURE	EST. LIFE
STRUCTURE	STRAIGHTED
STRUCTURE	TAMP
STRUCTURE	CLIMB AND INSP
STRUCTURE	REPLACE
STRUCTURE	NOTES
GUYS	ADD GUY
GUYS	TIGHTEN GUY
GUYS	REPLACE GUY
GUYS	RELOCATE GUY
GUYS	BOND GUY
GUYS	NOTES
CONDUCTOR – RESAG	PRIMARY
CONDUCTOR – RESAG	NEUTRAL
CONDUCTOR – RESAG	UNDERBUILD
CONDUCTOR – RESAG	OFFBUILD
CONDUCTOR – RESAG	SERVICE
CONDUCTOR	REPAIR STRAND
CONDUCTOR	RETIE
CONDUCTOR	NOTES
INSULATOR (REPLACE)	PIN TYPE
INSULATOR (REPLACE)	SUSP. TYPE
INSULATOR (REPLACE)	SECONDARY
INSULATOR (REPLACE)	SERVICE
INSULATOR (REPLACE)	NOTES
X-ARM	STRAIGHTEN
X-ARM	REPLACE
X-ARM	NOTES
RIGHT OF WAY	GROWTH IN FT.
RIGHT OF WAY	RECLEAR IN 20__
RIGHT OF WAY	GROWTH IN FT.
RIGHT OF WAY	RECLEAR IN 20__
RIGHT OF WAY	CUT DANGER TIMBER
RIGHT OF WAY	SIDE TRIM
RIGHT OF WAY	NOTES
MISCELLANEOUS	ARRESTER
MISCELLANEOUS	CUTOUT
MISCELLANEOUS	OCR
MISCELLANEOUS	TRANSFORMER
MISCELLANEOUS	LONG SECONDARY
MISCELLANEOUS	TIGHTEN HARDWARE
MISCELLANEOUS	RADIO NOISE
MISCELLANEOUS	NOTES
NOTES	ATTACH MORE SHEETS IF NECESSARY

81204-11_F34.EPS

Figure 34 Identification number.

using GPS. It eliminates the need to continually print and update paper map books and associated records. The software automatically updates the database when inspectors make entries.

With these systems, power company supervisors can set destination points, calculate routes, and guide work crews to maintenance locations. Inspectors can download all information necessary to perform an inspection. Work crew supervisors can download the same information when assigned to repair the faults discovered during inspection.

Data collection in the field uses technology such as GPS, bar codes, laser measurement, and **radio frequency identification (RFID) sensors**. GPS capability enables easy identification of geographical location, line, and area features for integration into the corporate database.

Online access may not be available in remote areas. In those cases, the data required for the inspection or maintenance is downloaded before going to the job site. Line workers enter changes on site and upload them later when communication access is available.

8.0.0 Environmental Concerns

Major maintenance projects on transmission lines, especially in wilderness areas, can have an impact on the environment. Such projects must consider the effects of brush clearing, changed migration routes for migratory animals, possible access by predators and humans along transmission right-of-ways, and disturbances of fish habitats at stream crossings.

Environmental planning is done by engineers when planning maintenance projects. Even so, work crews have to be aware of the effect their presence has on the environment and they are expected to follow all environmental regulations. They must stress minimal disturbance of the environment as part of their daily routine. Huge fines, lost time, and expensive rework can result from inattention to environmental issues.

8.1.0 Compliance Management Plan (CMP)

Power company engineers prepare environmental **Compliance Management Plans (CMP)** and apply mitigation efforts to minimize the impact of major maintenance projects. The CMP includes the following:

- Oversight of project environmental requirements
- Inspection, monitoring and reporting of the use of best management practices
- Technical review of preconstruction documents such as permits and technical specifications

The engineering department considers natural resources when creating a CMP for transmission line maintenance. Specific issues include vegetation clearing, erosion effects on nearby wildlife and plant habitats, consequences on wetlands from disturbance by equipment or from erosion, and the impact of new transmission structures. The assessment must take into consideration the potential for bird collisions with conductors, and possible electrocution of large birds of prey. The assessment has to identify any special use of defined habitats, such as wintering areas or special breeding habitats. The following factors are all considered when preparing the assessment:

- Migratory birds
- Bald and golden eagles
- Threatened and endangered species
- Impacts to common wildlife
- Water pollution
- Air pollution (equipment exhaust on local air quality)

- Erosion impacts (disturbed habitats, loss of land, and loss of water productivity)
- Impact on native plants and animals
- Impact on ecosystem diversity and structure
- Impact on sensitive vegetation
- Trash disposal
- Storage, handling, transport, and disposal of hazardous wastes
- Effects of **herbicides** on humans, plants, and wildlife
- Cultural and archaeological resources

8.2.0 Work Crew Concerns

The work crew at the job site has direct control over many factors that may affect the environment. Line workers must be careful that they do not cause any environmental damage, whether working on a major maintenance project or a routine job that takes only a few hours. They have to know enough to stop work and report to a higher authority if they become aware of maintenance efforts that have a negative effect on the environment.

8.2.1 Wetlands, Waterways, Lakes, and Streams

Wetland and other water sources can be affected during major maintenance such as reconductoring and transmission structure replacement. Erosion may affect physical conditions in the wetland, thereby altering the plant community composition.

Construction of access roads may reduce stream bank stability in areas of steep terrain, increasing the potential for sedimentation and reduction of water quality. For that reason, access roads must be routed away from lakes and streams. Wherever possible, avoid fording small streams with large construction equipment. Use existing bridges to avoid working close to streams and wetlands. Do not dig near streams, as banks can collapse. A collapsed bank can pollute water, harm fish habitats, and reduce or stop water flow.

If possible, perform ground inspections and repair activities during periods when surface disturbance would be minimal. Avoid rainy months. Power companies may have to limit construction to the winter months when soil and water are more likely to be frozen and vegetation is dormant (*Figure 35*). Yet, many wetlands never freeze, so work crews may have to use mats and wide-track vehicles when working in wetlands.

81204-11_F35.EPS

Figure 35 Winter work on frozen ground.

Heavy machinery can crush wetland vegetation and wetland soils. Wetland soils, especially those rich in peat, are easily compacted. Compacting such soil increases runoff, blocks flows, and greatly reduces the wetland's water-holding capacity. Construction and maintenance equipment driven across wetlands can stir up sediments that endanger fish and other aquatic life. Repeated driving of heavy construction equipment can wear away ground cover. Bare ground erodes during periods of heavy rain. The silt and sediment from the eroded area finds its way into waterways. Areas of exposed soil, especially on slopes and steep grades, must be reseeded.

Vehicles and construction equipment can introduce exotic plant species such as Purple Loosestrife. With few natural controls, these species may overwhelm native vegetation and destroy valuable wildlife habitat. Construction equipment must be washed after working in areas infested by known invasive or exotic species, and always before working in a sensitive area.

GOING GREEN

What are Wetlands?

First of all, wetlands aren't always wet, so a body of water doesn't necessarily qualify as wetlands. Wetlands are defined by a certain mixture of soil, water, and vegetation. The water may be below the surface all or part of the time. Wetlands are valuable because they provide natural filtration for water. For that reason, they are protected by law.

8.2.2 Woodlands

Work crews can reduce the impact to woodlands by avoiding routes that fragment large forested areas. Engineering may adjust transmission structure placement and span length to minimize the need for tree removal and trimming along forest edges. Tree and shrub species that reach heights of 12 to 15 feet can be allowed to grow within the right-of-way in sensitive woodlands.

Helicopters are often used for inspection and maintenance in remote areas in order to avoid the impact that vehicles might cause on forests and waterways. Because access to these sites may be infrequent, vehicles may have to blaze a trail through new-growth trees and brush to reach them. If vehicles are used, work crews may be required to wash them before entering protected woodlands to prevent the introduction of alien plants and plant diseases.

8.2.3 Clearing, Grading, and Vegetation Control

Major projects, such as reconductoring and transmission structure replacement, have the potential to expose large areas of soil, which increases the risk for soil erosion and sediment-laden storm water runoff. Engineers who plan large projects include steps to treat exposed areas that work crews will have to follow. No clearing and grading activities occur until sediment and erosion control measures are in place in accordance with the CMP. One of the most important measures is to place erosion barriers called siltation fences (*Figure 36*) in areas where ground cover is disturbed. Hay bales are used for the same purpose where there is a channel for water to run.

Depending on soil type and terrain, the CMP specifies whether to reseed or spread mulch. Most of the time reseeding is necessary (*Figure 37*). But in some soil types, spreading mulch on bare soil is all that is necessary to prevent erosion.

Vegetation removal and soil stabilization must be done in accordance with federal, state, and local guidelines along waterways. Existing vegetation should be cleared only where necessary. All excavations made by clearing activities must be backfilled. Disposal of cuttings and debris must be done at a facility designed and approved to handle such waste.

Undesirable vegetation in areas near public water supplies, open waters, wetlands, springs, wells, homes, or roadsides must be controlled by manual removal rather than by applying herbicides.

When clearing a right-of-way, leave a buffer strip of vegetation within a few feet of any body of water. Clearing of vegetation may affect water temperature and concentration of suspended solids from increased surface runoff. Do not apply herbicides in the buffer zone.

On Site

Kudzu

Kudzu is a leafy vine that is familiar to anyone living in the southeastern United States. It was introduced to this country in 1876 and was widely used as an ornamental. In the 1920s and 1930s, it was used as an animal feed. The Civilian Conservation Corps planted it extensively during the depression of the 1930s for erosion control, and farmers were paid to plant and cultivate it. The problem is that once it started to grow, no one could stop it. Herbicides do not kill it, and in fact can accelerate its growth. Today, kudzu can be found all over the southeastern United States. It climbs trees and other plant life, depriving them of sunlight, and killing them in the process. It is both a nuisance and a hazard for power line workers because it climbs utility poles and transmission structures. It is capable of conducting electricity, so it can cause electric shock as well as power outages.

81204-11_SA03.EPS

Figure 36 Erosion control methods.

Herbicides are generally applied by workers using handheld applicators. Federal or state authorities may grant an exception for machine application in unique circumstances. Herbicides must not be used during inclement weather or on excessively windy days.

8.2.4 Air Pollution

Most air pollution at a work site comes from equipment exhaust. Line crews must have some equipment running to perform their work, but must avoid unnecessary idling of engines. Engines not actually required for work activities should be turned off. If possible, equipment should be positioned so that the wind does not blow the exhaust toward homes and businesses.

Work crews have to comply with Clean Air Act requirements for mobile and stationary equipment. Depending on location, there may be controls for dust emissions that require sprinkling water on exposed earth.

8.2.5 Wildlife

The Endangered Species Act requires that any federally listed species and habitat must not be adversely affected during any maintenance activity. The company, and its workers, must ensure that line work and other maintenance activities do not disturb or harm any listed species or their habitat.

Figure 37 This area requires grading and reseeding when the work is complete.

Work crews have to be careful so that wildlife does not suffer harm from the modification of habitat. Animals that live in or near the right-of-way may have reduced growth and viability when a changed habitat alters their living and feeding habits. In addition, improper application of herbicides for right-of-way vegetation control may harm wildlife.

In remote areas that have nesting birds of prey, line workers may have to install devices on transmission structures and overhead ground wires to prevent perching. It may be necessary to place markers on the overhead ground wire to make the lines more visible to birds if the collision potential is high.

Maintenance along the right-of-way may disturb wildlife. It may be necessary to stop work when endangered, threatened, or sensitive species are mating and raising their young in a work area.

8.3.0 Trash Disposal

It is very easy to ignore trash and debris that falls on the ground. Always gather up trash and put it in the proper container for disposal. Remove and properly dispose of the waste off site at the end of each workday. Dispose of cuttings and debris in an approved facility designed to handle such waste. Never store trash within 100 feet of helicopter operations. On completion of construction, remove all remaining trash and debris from the site. Return all temporarily disturbed areas to their previous state, to the extent possible.

8.3.1 Hazardous Waste

The Toxic Substances Control Act regulates hazardous materials. Maintenance activities must be controlled to avoid sudden and accidental releases of pollutants and harmful chemicals. Approved procedures must be followed for the handling and disposal of hazardous waste.

Fuel tanks, lubricants, inhibitors, oily rags, trash, broken tools, and defective components removed from transmission structures are potential hazards. Examples of other materials that are found at work sites classified as hazardous wastes include organic solvents, paints, used oil, paint thinners, wastes that contain ignitable or corrosive materials, and wastes that contain certain toxic pollutants.

Any PCB waste from old transformers requires special handling to meet requirements for storage and disposal. Only licensed contractors may handle cleanup and disposal of such material.

Whenever herbicides or pesticides are being used, these basic rules should be followed:

- Transport herbicides and other hazardous materials in strong, leakproof containers that are properly stored and secured.
- Locate herbicide mixing sites a safe distance from lakes, ponds, rivers, streams, and wetlands. Do not mix herbicides in areas with soils that easily absorb and break down herbicides.
- Do not place tanks containing hazardous chemicals on a slope without taking steps to ensure the contents do not spill.
- Store all hazardous chemicals where vehicles cannot damage them. Keep hazardous waste away from areas subject to helicopter rotor wash.

8.3.2 Spill Reporting

There are emergency planning and reporting requirements for hazardous chemicals under the Emergency Planning and Community Right-to-Know Act (EPCRA) and for oil under the Oil Pollution Act. The exact requirements depend on the type of chemical handled.

NOTE

Any spill of a hazardous chemical that exceeds a specified quantity must be reported. If work crews spill or release more than this quantity at a maintenance site, the company must send a report to a designated local authority. The company environmental health and safety coordinator, or your supervisor, should have that information.

The EPCRA also requires that a Material Safety Data Sheet (MSDS) be available for all materials on the site that contain hazardous chemicals. An MSDS is available from the chemical supplier and on the Internet. All workers should read the MSDS for the materials they work with, both to protect the environment, and for their personal safety.

8.4.0 Blasting

Major transmission line maintenance projects may require blasting. If blasting is necessary, engineers assess each location with regard to apparatus or structures in the vicinity, and a determination is made of the suitability of that location for blasting. If a proposed blasting area is considered unsafe, excavation must be done by alternative means.

WARNING!

Blasting cannot be done within 200 feet of a buried pipeline or a building. Identify owners before blasting near structures and satisfy any concerns they may have.

Blasting operations should follow a blasting control plan designed by a specialist. Actual blasting work is normally conducted by a licensed contractor. The contractor prepares a blast pattern and shot design chart for each blast for review and approval by site supervisors. Normally, blasting occurs between the hours of 8 AM and 5 PM. Workers may have to assist in the installation of blast blankets where flying debris such as rock is anticipated.

SUMMARY

Transmission lines and the structures that support them can deteriorate over time. They can also suffer damage from a number of sources. Replacement of insulators damaged by vandalism and other causes is a common occurrence. Transmission lines are periodically inspected in order to identify problems, such as damaged insulators and conductors that require maintenance. Inspection and maintenance is generally performed by line workers who climb the transmission structures. These procedures may also be performed from helicopters, especially in remote areas.

While all standard safety precautions apply, there are additional precautions that must be followed because of the working heights, high voltages, and helicopter work. Critical among these precautions are the use of fall protection equipment; the need to establish equal potential zones whenever working on live lines; and the special precautions associated with working from helicopters.

Among the maintenance procedures performed on transmission lines are the replacement of insulators; reconductoring; installation and replacement of vibration isolators and line spacers; and washing of insulators to remove bird droppings, salt, and pollution. In some instances, lines are shut down so that maintenance can be performed. For safety reasons, it is important that line workers understand their company procedures for isolating lines.

Because inspection and maintenance work is done in and around forested areas and waterways, work crews must always consider the impact of their work on that environment. It is important to be conscious of regulations and company policies regarding sediment control, the use of herbicides, and the potential impact on wildlife, especially with regard to protected species.

Review Questions

1. The best method to prevent injuries due to falls is _____.
   a. using properly inspected fall-arrest equipment
   b. installing extra handholds while climbing
   c. training climbers to be nimble while ascending transmission structures
   d. using helicopters to eliminate climbing

2. A line worker should approach a helicopter from the _____.
   a. side staying within sight of the LSO
   b. front staying within sight of the pilot
   c. front or side within sight of the LSO
   d. front or side within sight of the pilot

3. Sharp points on a broken insulator result in _____.
   a. brittle failure
   b. stress failure
   c. corona discharge
   d. creepage

4. When washing insulators, wash from the _____.
   a. wide part to the narrow part so that water will concentrate at the narrow part
   b. narrow part to the wide part so that water will flow naturally to the wide part
   c. top to bottom to prevent water from dripping onto the cleaned section
   d. bottom to top to prevent water from dripping onto contamination

5. How does current flow affect conductors?
   a. It causes spacers on bundled conductors to deform.
   b. It welds compression splices together.
   c. It shortens conductor life due to elevated temperature.
   d. It increases conductor ampacity.

6. Higher current resulting from increased demand is likely to increase the sag in a transmission line.
   a. True
   b. False

7. Which of the following is a correct statement regarding the installation of OPGW lines?
   a. They can use the same travelers as power conductors.
   b. A swivel is generally used on the pulling line.
   c. Sagging requirements are much different from those of power conductors.
   d. OPGW lines are spliced in mid-span.

8. The type of conductor splice that uses a filler compound is the _____.
   a. compression splice
   b. automatic splice
   c. clamp splice
   d. helical rod splice

9. Bundled conductors are often tilted to compensate for _____.
   a. Aeolian vibration
   b. galloping vibration
   c. wake-induced vibration
   d. wind-induced sag

10. The maximum resistance of a tower ground should be _____.
    a. 150 ohms
    b. 100 ohms
    c. 50 ohms
    d. 10 ohms

11. Corrosion within 12 inches of the surface on guy anchors indicates _____.
    a. it is time to replace the epoxy coating on the guy anchor
    b. there is no problem at present, reinspect in five years
    c. there is likely to be further corrosion several feet down
    d. it is time to replace the zinc coating on the guy anchor

12. The danger of new construction near a transmission line right-of-way is that _____.
    a. it may cause erosion that will weaken a transmission structure footing
    b. tall buildings may be subject to corona
    c. tall buildings may touch the power line when sag increases on hot days
    d. transmission structure grounds may be disturbed, increasing shock hazards in the new buildings

13. A switching order is a _____.
    a. transmission line maintenance procedure
    b. set of engineering software instructions
    c. written document that specifies what switches to operate
    d. set of communications procedures that explain maintenance steps

14. Why would power companies limit construction efforts to the winter months?
    a. The frozen water prevents harm to fish and other aquatic life.
    b. The frozen soil and water are less likely to suffer harm from large equipment.
    c. The colder air is denser, which minimizes air pollution.
    d. The colder air carries sound better, which alerts wildlife to flee.

15. Herbicides should *not* be used _____.
    a. on excessively windy days
    b. on oak and maple trees
    c. in areas where line workers park equipment
    d. in areas under transmission lines subject to corona

Trade Terms Introduced in This Module

Clearance: See *isolation guarantee.*

Compliance Management Plan (CMP): A document that explains how a power company intends to minimize harm to the environment from transmission line maintenance.

Creepage: The conduction of electricity across the surface of a dielectric.

Creepage distance: The shortest distance between the two metal ends of an insulator when following the contours of the insulator.

Cycloaliphatic: A resin material similar to fiberglass used to manufacture suspension insulators. It is lightweight, very strong, and has the ability to repel water, which reduces the danger of flashover.

Dry band: A pollutant-contaminated area on an insulator that permits the flow of high current on the surface of the insulator. If the potential is sufficiently high, flashover will occur.

Flashover: A short circuit to ground caused by conductive contamination on the surface of an insulator.

Herbicide: A chemical that is spread on vegetation to kill unwanted plants. Some herbicides are selective, designed to kill specific types of plants while leaving other plants unharmed. Others are nonselective and kill all plants they contact.

Hydrophilic: The property of a surface that, when contaminated with pollutants, allows it to become saturated with moisture.

Hydrophobic: The property of a surface that, when contaminated with pollutants, enables it to repel water and reduce the danger of flashover.

Isolation guarantee: A document obtained before starting work when it is necessary to de-energize a circuit to enable work to be safely performed. It is intended to protect workers from harm on a circuit that presents a hazard.

Landing signal operator (LSO): A person who directs the landings, takeoffs, and hoist operations of helicopters.

Overhead ground wires (OHGW): Multiple grounded wires placed above phase conductors for the purpose of intercepting direct lightning strikes in order to protect the phase conductors from the direct strikes.

Radio frequency identification (RFID) sensor: An electronic device that transmits a coded signal to a receiver that decodes the signal.

Supervisory Control and Data Acquisition (SCADA): A central computer system that collects data from sensors. The central computer system processes the data and transmits control signals to operate remote equipment.

Spalling: Concrete deterioration in which the outer section of stone, brick, or mortar breaks away in a localized area. The cause is usually the freezing and thawing cycles of moisture trapped within the concrete.

Switching order: A written set of instructions that details what switches to operate and in what sequence, in order for line crews to work safely.

Additional Resources

This module presents thorough resources for task training. The following resource material is suggested for further study.

Electric Power Transmission Systems, 2nd Edition, Upper Saddle River, NJ: Prentice Hall.

Figure Credits

KT Power, Inc., Figures 1, 7, 8, 12–16, 18, 20, 23, 24, 26, 28, 29, 31, 35–37, and SA01

HD Electric Company, Figures 2 and 9

MD Helicopters, Figure 3

Terry Williams, Figures 4 and 25

Chesapeake Bay Helicopters/CBH Aviation, Inc., Figure 5

U.S. Department of Transportation, Federal Aviation Administration, Figure 6

Classic Connectors, Inc., Figures 10, 17, and 21

Tyco Electronics Energy Division, SA02

Simplex Manufacturing, Figure 11

Power Transmission Solutions, Figure 19

Wilson Construction Company, Figure 22

Syntronics, Figure 27

Topaz Publications, Inc., Figures 30, 32, 34, and SA03

Electromatic Equipment Co., Inc., Figure 33

NCCER CURRICULA — USER UPDATE

NCCER makes every effort to keep its textbooks up-to-date and free of technical errors. We appreciate your help in this process. If you find an error, a typographical mistake, or an inaccuracy in NCCER's curricula, please fill out this form (or a photocopy), or complete the online form at **www.nccer.org/olf**. Be sure to include the exact module ID number, page number, a detailed description, and your recommended correction. Your input will be brought to the attention of the Authoring Team. Thank you for your assistance.

Instructors – If you have an idea for improving this textbook, or have found that additional materials were necessary to teach this module effectively, please let us know so that we may present your suggestions to the Authoring Team.

NCCER Product Development and Revision

13614 Progress Blvd., Alachua, FL 32615

Email: curriculum@nccer.org
Online: www.nccer.org/olf

❑ Trainee Guide ❑ AIG ❑ Exam ❑ PowerPoints Other ______________________________

Craft / Level: Copyright Date:

Module ID Number / Title:

Section Number(s):

Description:

Recommended Correction:

Your Name:

Address:

Email: Phone:

Glossary

Apex: The point of sling attachment to a crane hook.

Armor rods: Metal rods that are placed around a conductor as it passes through the suspension clamp. They are used to stiffen the conductor to prevent bending and flexing. They are often called line guards.

Articulated running board: A device used when pulling multiple (bundled) conductors through a stringing block. The running board allows all conductors to feed through the stringing blocks without becoming tangled.

Blind lift: A lift where the crane operator(s) or a helicopter pilot cannot observe all or part of the load movement.

Bullwheel tensioner: A machine that contains reels of conductor to be installed. As the conductor is unreeled, the bullwheel tensioner along with the pulling machine, maintains the correct pulling tension.

Capacitance: The storage of electricity in a capacitor; capacitance produces an opposition to voltage change. The unit of measurement for capacitance is the farad (F) or microfarad (μF).

Circuit: Any set of three-phase transmission lines used on a structure.

Clearance: See *isolation guarantee*.

Compliance Management Plan (CMP): A document that explains how a power company intends to minimize harm to the environment from transmission line maintenance.

Compression sleeve: A type of conductor splice that is used to connect two conductors. It is squeezed around the conductors with a hydraulic crimping tool.

Corona ring: A doughnut-shaped device that is often installed on insulators on high-voltage lines to reduce the effects of corona.

Creepage: The conduction of electricity across the surface of a dielectric.

Creepage distance: The shortest distance between the two metal ends of an insulator when following the contours of the insulator.

Cycloaliphatic: A resin material similar to fiberglass used to manufacture suspension insulators. It is lightweight, very strong, and has the ability to repel water, which reduces the danger of flashover.

D/d ratio: The relationship between the diameter of the surface (D) that a wire rope sling is around and the diameter of the sling wire rope (d).

Dead-end structure: Normally used at the endpoints of specified lengths of line conductors after a specified number of spans between suspension structures.

Deviation/tension structure: Used to sharply change the direction of transmission lines horizontally or on inclines.

Double Kellems grip: A double-ended wire mesh device that is used to make a temporary splice between conductors that are being pulled through stringing blocks. A double Kellems grip is used because some permanent splices cannot be pulled through stringing blocks.

Drum-style pulling machine: A powered pulling machine on which the pulling rope is wound as it pulls a conductor through stringing blocks. It is used with a bullwheel tensioner to help maintain proper line tension during stringing.

Dry band: A pollutant-contaminated area on an insulator that permits the flow of high current on the surface of the insulator. If the potential is sufficiently high, flashover will occur.

Dynamometer: A device used to measure tension on conductors as they are being pulled through stringing blocks.

Electromagnetic induction: It is the principle that allows a transformer to function. When the fluctuating electromagnetic field around a conductor that carries alternating current cuts through a nearby conductor, an alternating current is induced in the other conductor.

Electrostatic induction: It is the principle that allows a charged object to induce a static charge in a nearby object. When an object containing a charge is brought next to an object with no charge, it draws ions with the opposite charge to the surface and creates a static charge between the two objects.

Equal potential zone: A work area in which the potential (voltage) on all surfaces that can be reached or touched by a worker are equal. The zone is established by having all surfaces grounded at a common point with temporary ground jumpers.

Flashover: A short circuit to ground caused by conductive contamination on the surface of an insulator.

Frequency: The number of cycles an alternating electric current, sound wave, or vibrating object undergoes per second.

Full-tension splice: A type of splice meant to restore strength and conductivity to a conductor. Compression splice sleeves are widely used.

Ground mat: A conductive mat placed on the ground to establish an equal potential zone. The mat is typically used to eliminate touch and step potential hazards.

Herbicide: A chemical that is spread on vegetation to kill unwanted plants. Some herbicides are selective, designed to kill specific types of plants while leaving other plants unharmed. Others are nonselective and kill all plants they contact.

Hertz (Hz): A unit of frequency; one hertz equals one cycle per second.

Hydrophilic: The property of a surface that, when contaminated with pollutants, allows it to become saturated with moisture.

Hydrophobic: The property of a surface that, when contaminated with pollutants, enables it to repel water and reduce the danger of flashover.

Impedance: The opposition to current flow in an AC circuit; impedance includes resistance (R), capacitive reactance (X_C), and inductive reactance (X_L). Impedance is measured in ohms.

Inductance: The creation of a voltage due to a time-varying current; also, the opposition to current change, causing current changes to lag behind voltage changes. The unit of measure for inductance is the henry (H).

Isolation guarantee: A document obtained before starting work when it is necessary to de-energize a circuit to enable work to be safely performed. It is intended to protect workers from harm on a circuit that presents a hazard.

Kellems grip: A wire mesh grip used to attach a pulling rope to the conductor so that it can be pulled through stringing blocks. Other names for a Kellems grip include basket grip, sock, and Chinese finger.

Landing signal operator (LSO): A person who directs the landings, takeoffs, and hoist operations of helicopters.

Lattice steel tower (LST): A transmission structure made up of numerous steel strips and angles.

Micro: Prefix designating one-millionth of a unit. For example, one microfarad is one-millionth of a farad.

Minimum approach distance: A working distance to energized objects. Outside the minimum approach distance, no rubber gloves or sleeves are required, and conducting materials and tools can be handled. At distances less than the minimum, rubber gloves and sleeves must be worn and protective shields and nonconducting tools must be used. The minimum approach distance varies. As voltage increases, so does distance.

Optical ground wire (OPGW): A type of shield wire used to protect transmission structures from lightning. It has outer layers of steel and aluminum for strength and conductivity and an inner core that contains optical fibers. The outer layers act as the ground, while the inner optical fibers are used for the utility's voice and data signals.

Overhead ground wires (OHGW): Multiple grounded wires placed above phase conductors for the purpose of intercepting direct lightning strikes in order to protect the phase conductors from the direct strikes.

Peak voltage: The peak value of a sinusoidally varying (cyclical) voltage or current is equal to the root-mean-square (rms) value multiplied by the square root of two (1.414). AC voltages are usually expressed as rms values; that is, 120 volts, 208 volts, 240 volts, 277 volts, 480 volts, etc., are all rms values. The peak voltage, however, differs. For example, the peak value of 120 volts (rms) is actually $120 \times 1.414 = 169.68$ volts.

Pilot line: A lightweight line used to draw a pulling rope through stringing blocks.

Radio frequency identification (RFID) sensor: An electronic device that transmits a coded signal to a receiver that decodes the signal.

Reactance: The imaginary part of impedance. Also, the opposition to alternating current (AC) due to capacitance (X_C) and/or inductance (X_L).

Rescue stretcher: A stretcher designed to lower badly injured and/or unconscious workers from high places.

Resonance: A condition reached in an electrical circuit when the inductive reactance neutralizes the capacitance reactance, leaving ohmic resistance as the only opposition to the flow of current.

Root-mean-square (rms): The square root of the average of the square of the function taken throughout the period. The rms value of a sinusoidally varying voltage or current is the effective value of the voltage or current.

Running ground: A temporary ground jumper equipped with rollers that keeps the conductor grounded while it is being pulled through stringing blocks.

Sag scope: An optical device similar to a telescopic sight on a rifle. It is mounted on a transmission structure and used to sight along power lines to check for correct sag.

Self-inductance: A magnetic field induced in the conductor carrying the current.

Spacer cart: A lightweight cart that can travel between transmission structures while suspended from bundled conductors. Workers riding in the cart can install spacers or do other tasks.

Spalling: Concrete deterioration in which the outer section of stone, brick, or mortar breaks away in a localized area. The cause is usually the freezing and thawing cycles of moisture trapped within the concrete.

Step potential: The potential (voltage) that exists between a person's feet when that person stands near an energized grounded conductor such as a transmission structure leg. Eliminate step potential by setting up an equal potential zone.

Stringing block: A pulley attached to insulators through which conductors are pulled while being installed. They are often called travelers.

Supervisory Control and Data Acquisition (SCADA): A central computer system that collects data from sensors. The central computer system processes the data and transmits control signals to operate remote equipment.

Suspension clamp: A clamp hung below the insulator or yoke plate that supports and grips the conductor after it has been sagged and dead-ended.

Suspension structure: Used along relatively straight and level runs of transmission lines. They are also used for long shallow turns.

Switching order: A written set of instructions that details what switches to operate and in what sequence, in order for line crews to work safely.

Tension stringing method: A method used to string conductors between transmission structures. The conductor is pulled under tension over stringing blocks on each transmission structure. The method keeps conductors off the ground to avoid damage. It is the preferred method for installing new conductors.

Touch potential: The potential (voltage) that exists between a worker and an energized grounded conductor such as a tower leg. Eliminate touch potential by setting up an equal potential zone.

Transmission structure: Used to support overhead lines (conductors) for electric power transmission.

Tubular steel pole (TSP): A transmission structure that uses round steel members.

Yoke plate: A plate mounted directly below a suspended insulator(s) on which one or more suspension clamps can be attached.

Index

D

E

K

L

M

N

T

U

V

W

X

Y